— Frank Sawyer —

KEEPER *of the* STREAM
The life of a river and its trout fishery

For Carl,
who has been my wise
keeper on more than one occasion!

C. t

Christmas, 1985

— Frank Sawyer —

KEEPER *of the* STREAM
The life of a river and its trout fishery

*with a new introduction by Sidney Vines
and illustrations by Charles Jardine*

London
GEORGE ALLEN & UNWIN
Boston   Sydney

First published in 1952 by A. & C. Black
This new edition published in 1985 by George Allen & Unwin
© Illustrations by Charles Jardine, 1985

**George Allen & Unwin (Publishers) Ltd,
40 Museum Street, London WC1A 1LU, UK**

George Allen & Unwin (Publishers) Ltd,
Park Lane, Hemel Hempstead, Herts HP2 4TE, UK

Allen & Unwin Inc.,
9 Winchester Terrace, Winchester, Mass 01890, USA

George Allen & Unwin Australia Pty Ltd,
8 Napier Street, North Sydney, NSW 2060, Australia

ISBN 0 04 799026 0

Printed in Great Britain
by Anchor Brendon, Ltd., Tiptree, Essex

CONTENTS

INTRODUCTION

FRANK SAWYER was born in 1906 in a cottage at Bulford in the valley of the upper Avon in Wiltshire, so close to the river that, as he wrote himself, the sound of it cascading through the hatches must have been one of the first sounds he ever heard. In 1928, he became keeper of the Services Dry Fly Fishing Association water—the six and a half miles of the Avon which stretches upstream from Bulford. Thus, when *Keeper of the Stream* was first published in 1952 he had lived 46 years by the river, and had 24 years experience of caring for it as keeper. The essence of what he had learnt is distilled into this book.

His main task was to provide sport for the members of the Association: this meant seeing that there was a good stock of trout, with as many as possible being wild fish spawning naturally in the river and sidestreams, and also food for the fish in the form of aquatic insects to which they would rise. But this book shows how much wider were Sawyer's interests. Everything about the river valley fascinated him, from large animals and birds like otters and herons to tiny midge larvae, which are food for trout fry. When mayfly hatched from the river, he followed them across the meadows to see the whole process. He observed specimens under the microscope he had managed to acquire, and he made careful notes. Even at the end of his life—he died in 1980— his eyes would light up with interest and excitement when fishing prospects for the next season were discussed, just as they had done more than fifty years before, when he began as keeper.

His career as a writer began late and might never have begun at all. During the Second World War, he met a retired judge called Sir Grimwood Mears who was at the time a widower living in a hotel at Amesbury, about three miles downstream from Bulford. Sir Grimwood was a keen fisherman, who spent much time by the Avon, and inevitably he got to know Sawyer, a War Reserve policeman stationed at Netheravon. Sitting on a bench during lulls in the fishing, they would naturally discuss current topics in the fishing world, one of which was the question of nymph fishing. Sawyer had given much thought to this, and read the works of G. E. M. Skues with admiration, but did not agree with everything Skues had written about the behaviour of the natural nymph. Skues was a fine observer of nature, but so was Sawyer, and Sawyer had a great advantage. Whereas Skues visited the river in his leisure hours (he was a solicitor), Sawyer lived and worked by it and—most importantly—for a good deal of the time he was *in* it.

Up to his waist in water cutting weeds or repairing banks, he would peer into the depths with his extraordinary ability to see below the surface, and observe what he called the 'active, or restless' nymphs. From these observations he evolved what became known as the Netheravon style of nymph fishing, and which was to captivate such good judges as Charles Ritz and Howard Marshall. His first description of the theory is in Chapter 9 of this book.

As Sir Grimwood got to know Sawyer, and heard his ideas, he realised that he had found a man of exceptional promise and talent, but who because of his shyness and lack of education needed help. This, Sir Grimwood determined to give, and he was the ideal man—a master of clear, concise English and well acquainted with the leading figures in the fishing world. In 1945, he arranged for Sawyer to meet Skues, then in his nineties and living in retirement in a hotel at Wilton. As so often, the extraordinary bond created by fishing worked its magic. Skues, the greatest authority of his day, and the humble river keeper, struck up a rapport which lasted until Skues' death in 1949.

They had a long correspondence, mostly about entomology, and Sawyer carefully preserved Skues' letters. They are written in a tiny yet meticulous hand, and cannot be read without a magnifying glass. In some of them, Skues shows himself concerned with Sawyer's future. He was a great gentleman, not resenting in the least that the young keeper dared to question some of his theories, but seeing only that here was a man with something worthwhile to say. He introduced Sawyer to Black's, the publishers, and to the editors of various angling magazines, among whom was R. L. Marston, then editing the *Fishing Gazette*.

The Marston family had made the *Fishing Gazette* an institution. R. B. Marston (father of R. L.) edited it in Victorian times. He was a highly influential figure who is now generally recognised as the founder of the Flyfishers Club. His sons R. L. and E. W. took over after his death in 1927, and when E. W. died R. L. continued alone. With Mr John Eastwood he was a moving spirit in the founding of the Anglers Co-operative Association. In 1936, he published an article by a teenage writer called Richard Walker, and encouraged him to continue. In 1957, a few months before his death, he came with his daughter Patricia (now Mrs Richard Walker) to Netheravon and fished with Sawyer. Patricia caught her first trout at Netheravon, and after her father's death, continued to edit the *Fishing Gazette* successfully for five years until she gave it up for family reasons.

By 1951, Sawyer had written over a hundred articles for the *Fishing Gazette*. One day, Sir Grimwood showed a file of cuttings of the articles to his friend Mr Wilson Stephens, then Editor of *The Field*. Mr Stephens, struck by their consistency and quality, saw at once the potential for a book. He mentioned the idea to Sir Grimwood, who reacted favourably. Mr Stephens then approached Black's, the publishers, who were interested. R. L. Marston waived copyright, and Sir Grimwood conducted long drawn out negotiations with Black's on Sawyer's behalf. These at last came to fruition, and Mr Stephens then arranged the material into the book *Keeper of the Stream*.

It established the name of Frank Sawyer as a master of his subject, and there is a sense in which it is unique in the whole rich heritage of fishing literature. Only one name among river-keepers can be compared with Sawyer as an innovator—William Lunn of Stockbridge, immortalised by the spent flies 'Lunn's Particular' and 'Lunn's Yellow Boy'. His life and work was recorded by John Waller Hills in his classic *River Keeper*. The present book is of the same genre, but with the inestimable advantage for posterity that we have it all at first hand.

Where nomenclature has changed—such as hook sizes—I have added a footnote. Otherwise the text is as Sawyer wrote it. But I think readers should know that in three respects his views, as expressed in this book, changed as he grew older. In Chapter 14 he sees the dabchick only as a pest because it feeds, among other things, on trout fry. Twenty years later he put more emphasis on the dabchick as an indication of the purity of the water. 'Those who own trout waters' he wrote 'should be thankful that their fishery is considered by this little bird to be worth living on.'

His views on herons changed in much the same way. In Chapter 14 he gives a thrilling account of watching a hunting heron which he eventually shot. Herons are now, of course, protected by law, but in any case the older Sawyer would have tolerated it. He wrote 'I am not one of those who think fishing is all for the human. We should be prepared to share with those who have the greater right.' These are noble words, which all anglers would do well to heed.

In 1952, he had little respect for the grayling, regarding it as vermin. At that time, grayling in the Avon far outnumbered the trout and they had to be drastically reduced. In later years, when the balance was better, he could see that the grayling had its uses. It provided good fishing in October when the trout were getting near to spawning.

Several times in the early seventies, he and Mrs Sawyer were invited to Austria as guests of HRH The Duke of Bavaria. The intention was for him to fish the river Salza for grayling. He had magnificent sport with fish of more than two pounds in excellent condition. He later wrote: 'In future I shall think more kindly of them on the Avon, where they must still be controlled, for I realise that these should be the fish of the cold, clear, waters of the mountains, and they are far from home.'

Since I first read this book, more than twenty years ago, one passage has always stayed in my mind. It is in Chapter 10, where Sawyer recounts how he met, after many years, a man who he had first known as a young Captain but had now become a General. His face was "aged and lined with worry." Sawyer muses: 'Even as he had risen in his profession, I felt I had risen in mine . . . I began to wonder which of us was the more content.'

Most of us go through life trying to better ourselves. We seek promotion, we move about the country or overseas, we expand our responsibilities. We seek the bubble, reputation. Sawyer's life was completely different. Once appointed as keeper, at the age of 21, of the Services Dry Fly Fishing Association water, all his worldly ambitions had been fulfilled. His life's work would be to care for the river and for all the wild creatures of the valley who lived by and in it: and in studying them he was able to lift for us a corner of the veil beneath which nature keeps her secrets. This gave him a deep contentment which many, with a much greater share of the world's goods, could not but envy. This is a book about a man fulfilled.

SIDNEY VINES
Salisbury,
Wiltshire,
England.

1   LIFE IN THE VALLEY

JANUARY brings the start of the river keeper's year—the start of new seasons which may bring to him pleasure or disappointment, seasons which may prove the result of his vigilance during the past four or five years. His patience and perseverance may this year be rewarded and, in any case, he can look forward to the next twelve months with a certain eagerness or anticipation, for he is happy in the thought that at some time during the seasons he may learn something new, and see the unfolding of more of Nature's many secrets.

A river keeper's job is to assist Mother Nature, and Mother Nature can be capricious. She has to be studied, and studied thoroughly before she can be assisted, or she strongly resents interference. To be successful in producing either trout or insects, or both, it is first necessary to know something of the habits and life cycles of these creatures, to learn of their food, their environment, and their enemies. Here is the most interesting part of a river keeper's life.

In rivers, as elsewhere, one thing preys upon another, forming a vast cycle in which one creature is the food of

1

something else. In rivers usually the smaller animals are the food of those which are larger. So it is necessary to start at the beginning with the first living creatures and, putting this first life at the bottom of the ladder, work patiently towards the topmost rung. It is the tiny things—the young—which need the most assistance, for life in every instance commences in a very humble way.

In this manner only can Nature be assisted. If suitable environment is made for the smaller animals, then it will also be suitable for those that are larger, and so on up the scale. If one is to keep trout in a river without feeding them artificially, then that river must produce trout food, and if one is to produce successfully the insect life which provides a means of existence for the trout, one must first make sure that for the insects too, there is food.

The sport of dry-fly and nymph fishing for trout depends largely on fly life, and a large percentage of flies are bred in the river. So it is also necessary to assist in the production of the species of fly which are of most value from the fisherman's point of view; flies which will, in the completion of their life cycle, be present in and on the surface of the water to form an attraction which tempts the trout to rise to the surface to feed on them.

Each season of the year has its attraction; as the weeks change to months so change the character and appearance of the river, and the life of the river and riverside creatures. The river keeper need never be lonely, for he learns to understand and appreciate the wild life which are his constant companions, and as he sees the fruits of his labours all around him he gets a certain satisfaction.

Though neither are welcome, I much prefer floods to drought. I am sure far more harm is done to a fishery through lack of running water than by having too much. People often ask me what happens to the fish in time of flood. I think they imagine that they get swept out on to the meadows and are there stranded beyond hope of recovery. But though floods can do quite a lot of damage in other ways, I have never yet experienced any great loss of big trout, or other fish. Where a river is man made, trouble is much more likely to occur than where the streams have remained natural. I am never much concerned with fish after they are past the yearling stage—my chief worry is with the food of fish.

In time of flood wild fish are very capable of looking after themselves. Coarse fish usually remain in the river course somewhere near their usual haunts. They may move into the mouth of some adjacent sidestream, but generally are to be found taking shelter in eddies and backwaters near the river banks. Trout and grayling may work up-stream to the nearest shallow—trout to find a sanctuary under a bank, a tree-root or in a weed-bed, while grayling find a position where a hole in the gravel of the shallow gives a little seclusion from the racing water. It is seldom indeed that they range out over the river banks.

On the other hand the lesser creatures, though having a certain instinct of self-preservation, have not the strength to carry it into effect like fish. A sudden rise of water, especially if it is running fast, will sweep many millions of tiny living things over the river banks and out into the meadows beyond. Some may be entangled in weed growths which have been torn from the river bed with the force of the current; others may be forced to seek different environment because the added pressure of water, or its dirty condition, is too much for their respiratory powers, or for their frail bodies to withstand.

But even though they are swept out of the river bed, this does not necessarily mean that they are lost, for all is well if the flood water subsides gradually. The little creatures can then sense the dropping of the level, and will move with the receding water until eventually they regain the river itself, or some tributary which links with it—at all events, a place of safety. Should a quick drop in the level take place, then it needs little imagination to know what can happen. With the sudden draining off of the water, many millions of these little organisms become trapped in pools about the low-lying ground. There they

3

remain until the last drop of water has soaked away, and they are lost forever.

Excepting the Mayfly larvæ, and the river worms, which are secure deep in the river bed, and perhaps such fly larvæ as the Turkey Brown, Yellow May Dun and Claret Dun, which have a habit of clinging fast to the underpart of stones, few of the lesser creatures of a south country stream remain safe in the river course in time of flood. However, it is a natural event for a river to overflow, and it is also natural for the waters to drain away. All the wild life in a river has a certain instinct which guides and preserves it, and only when Man interferes with the mad idea of reducing a flood level in the shortest possible time is any great loss experienced.

It distresses me to see the flooding of a valley for with it sometimes comes tragedy, not so much perhaps for the aquatic creatures which live in the river and streams, but for the thousand others which are part of the valley life. I think sudden flooding is a thing for which Nature did not prepare all her family, and a quick rising and overflowing of a river finds some of them at a disadvantage. A few years ago a cloud-burst sent a wall of water racing down the valley to the sea—a wall which, as it advanced, spread far out on either side of the river course, to cover all but the highest parts of the low-lying meadows.

I went down the valley that day and the tragedy I saw I shall never forget. Though I tried, I was powerless to help to avert it. It all happened so suddenly. Water swept over the river banks of reach after reach, and flowed out across the meadows like a tidal wave. As the vegetation of the valley was covered, so the surface of the water became alive with panic-stricken, struggling creatures. Moles, shrews and fieldmice swam side by side, moving from one dry patch of land to another amidst hordes of insects of many descriptions that were milling about and drowning. Field voles and water voles came bobbing up from their sodden nests in search of refuge. Some clambered up trees —some scrambled on to driftwood or other flotsam, and were swept away down the valley—poor, shivering little animals whose homes were lost for ever.

At one place the centre part of a foot-bridge rose above the swirling waters, and though a hundred yards away I could see movements of life upon it. I walked through water nearly to the tops of my waders, and then could see what they were. Two water voles dived as I got nearer and swam to a nearby tree, but huddled together on the wooden bridge were three moles, seven fieldmice and two little pigmy shrews, all with their fur sodden, and shivering as much with fright as with cold.

I had a sack with me, and none of the little animals made a move to escape as I took them by their tails and dropped them into it. I turned them all loose amongst the tufts of grass on the high ground and in a moment all had disappeared. The three moles bored into the ground as though their lives depended on their getting beneath the surface. Perhaps they did. I have heard experts say that a mole cannot live more than a few minutes without food. I had much proof that day that they can. Perhaps these three were more hardy than some, for I learned afterwards that a friend of mine had seen them on the bridge an hour earlier, and he thought the pigmy shrews were two of their babies.

Moles were my chief interest that day. At one place, I could see the ground heaving at several points on a small mound about the size of a kitchen table. It was sixty yards away from me, and the water surrounding it was too deep for me to wade. As the water rose I could see the movements gradually concentrating towards the centre, then first one, then another, and finally four little black bodies appeared on the short grass. Panic-stricken they ran here and there, to stop and turn at the edges of the water, and then the first made up its mind to swim.

People say moles are short-sighted; perhaps they are. Some say they are blind, but it seems to me that Nature has provided them with an uncanny sense of direction. Each of these moles swam to where I was standing on the high ground and which for them was the only place of safety. All were exhausted as they landed but, with scarcely a pause, each bored down into the ground and, with the soil moving above their heaving shoulders, began tunnelling towards still higher land.

At one place down the valley a heron was at work. Herons make a prey of many of the little mammals of the riverside, and this one was quartering one of the meadows like a barn owl. Every now and then it sighted a swimming mouse, or vole and, dropping swiftly into the water beside it, with a quick stroke of its beak put an end to what chance the little animal had of escape. I suppose it is Nature, but I felt he was taking a very unfair advantage.

Man, when interfering with Nature, often causes up-heavals in the lives of wild creatures. One day I released a head of water to wash clear the cut weed in a lower reach. During the previous three weeks the sluices had been regulated to maintain a constant high level in the upper reach and when this impounded water was set free the river down-stream quickly rose bank high. As I stood on a bridge just below the sluice gates, a movement at the edge of the river caught my eye. I thought a fish had risen and watched the spot to see if the movement was repeated. A few seconds passed, then up through the water came a water vole. Something about its head looked unusual. I looked closer as it came swimming down towards me and landed at the base of an old withy pollard. Then I could see it had a young one in its mouth. Quickly it scrambled into a hole amongst the roots.

After a few moments the vole came out without the baby and, running rapidly along the bank, dived into the water at the spot where I had seen its first appearance. Soon it was up again with a second baby. The same route as before was taken to the tree-root. Then twice more the procedure was repeated. After waiting about ten minutes for the little animal to reappear, I decided she had taken in the last of the family. She was only just in time. The rising water had by then spread over the river bank and all but the tall grass at the river edge was submerged.

Going quietly to the tree I peered down into the hollow of its trunk. I could hear rustlings and tiny squeaks and, as my eyes became accustomed to the dim light, I could see the mother vole crouched amongst a litter of dried grass and rushes, busily pulling the material around her. As she worked so now and then she exposed one of the young ones as they wriggled about beneath her. I watched with interest, but eyes other than mine were also watching for there, humped up on a ledge, was a male vole. It was his couch she had taken over for her family in time of need. He made no attempt to help or to interfere, but his senses were keenly alert. I suppose I made some sound or quick movement. Like a flash he plopped into the water beneath him while she, for a moment, looked straight into my eyes. She froze and then leapt out of the nest to follow the male into the water. There exposed in the hollow nest were four little pink bodies trying to nestle into each other for warmth.

Poor little creatures, I thought. They were only a day or two old, and to be dragged rudely through water at this age they were indeed starting early to be water voles. I moved away quietly. It had been my fault. My lifting of the sluices had upset the calculations of this little family—the flooding was something for which they had been unprepared. Then as I continued with my thoughts, I began to wonder if this family was the only one that would suffer.

As I walked down-stream, soon there was evidence of other disruptions in the life of the wild creatures. There, sweeping down the river on a mass of cut weed, was a solitary, freshly hatched moorhen chick and, swimming with the current, clucking with agitation, was the mother bird doing her utmost to get the chick to obey her commands. I was relieved when the tiny black ball of fluff raced to the edge of the weeds and entered a rush-bed at the side of the river. The frightened cheeps of the little chick were getting fainter when I saw the nest in which it had been hatched. But no longer would it be used, for now it was crushed and broken where it had been swept beneath the overhanging branches of a tree.

I made a short cut to the hatches below so that I could open them and let the drifting weeds pass on down the river. On one big mass was another nest with a moorhen still sitting on her clutch of eggs. To me it seemed that she was quite unaware that she was adrift and not until the weeds were being sucked down through the open hatch did she realise the danger. Then abandoning nest and eggs with a series of alarm clucks, she fled to a nearby bed of sedge. Soon after this the nest of a dabchick, in which four eggs peeped from beneath their thin covering of weeds, passed through the hatchway into the vortex below. That was the fourth incident in less than an hour. Possibly, there were others of which, blissfully, I was ignorant.

Now these creatures are all weather-wise. They know when to expect natural events much better than we do. They had built their nests in which to have their young, safe in the knowledge that the river was not likely to rise through the effects of rain or other elements. But for my interference— my action in artificially raising the water level—there would have been nothing to endanger their family life.

About half an hour later I went back up-stream and with rather a guilty conscience I peered into the withy tree to know how the little voles were faring. But all I could see was the hollow where they had rested. That fleeting glance the mother had given to me had been sufficient. She thought she recognised an enemy and I felt strangely sad. I had unwittingly given her cause for further alarm and once more she had moved her family.

I hate to see life wasted, especially if it is life I try to encourage, and when my own thoughtlessness brings death in its train it is even harder to think about. One year I had a very successful season in the hatchery and by mid-April we had planted out some 40,000 healthy trout fry. My assistants and I had spent many hours ensuring that these little fish were widely distributed, both in the river and in sidestreams; in places along the edges of the water, where experience and study had told me they were likely to thrive. As I wandered along the banks for a few days afterwards there was plenty of evidence that the little trout had settled comfortably in their various homes. It gave me great pleasure to watch them as they poised in the water, or darted here and there for a few inches, to secure drifting insect life. However, towards the end of April, torrential rains swept the valley and surrounding downs; surface water poured into the river from the higher land and, as many of the controls were closed, the river quickly rose above its banks, and spread out into the meadows.

I thought of all those happy little trout I had seen feeding so eagerly a couple of days previously, and without thinking of the consequences I hastened round to all the controls and quickly released the flooding waters. I ought to have known better than to do such a thing but, as I have said, I just did not think. Knowing as I did that the little trout would move out of the river and sidestreams as the pressure of water increased, my one intention was to prevent them being stranded. By releasing the pent-up flood water so suddenly, I did the very thing I had hoped to prevent, for the flooding water quickly receded—in fact it dropped in level so fast that, in an hour, many of the little trout were stranded without a hope of regaining the river.

That same afternoon I had a feeling I had made a mistake, and wandered along a footpath, which was close to the river's edge, by a favourite shallow—a place where we had planted hundreds of trout fry, and where I had noticed they had become established. In various low parts of the path, little puddles of water were still trapped, and I searched them closely. In each were some of my trout fry. I found others still alive, but hopelessly trapped, in water-filled hoof-marks made by cattle.

What had happened was that when the little fish sensed or found that the water was rising, they followed a natural impulse to move to the edges of the river. They had wriggled through the narrow fringe of vegetation on the river bank, and had found comparatively clear, quiet water along the footpath. Those I found I transferred to the river, but I hate to think of the hundreds, probably thousands, which were stranded in other places where it was next to impossible to see them. For one must remember these little trout were scarcely 1 in. long, and of a colour which Nature has provided to merge with the vegetation.

However, I learned a lesson I have never forgotten, for had I left the little fish to the care of Nature—to their own instinct of self-preservation—many of them would have lived, for, though it is not unnatural for a river to rise quickly and flood the banks in its natural process, the level is seldom reduced suddenly, and while the floods are gently subsiding, the trout and insect life gradually move back once more to safety.

2   THE WATER

FEW rivers run very far down a valley without having a little brook, sidestream or minor river coming to join them from some low-lying part of the countryside. These little waters are usually fed by springs all along their course, and in all but very dry seasons, have a continual supply of pure water flowing down them. We have many of these little streams on the Upper Avon and I consider them to be very valuable to the fishery. I think of them as nurseries for the main river—nurseries, not only for trout, but for the multitude of life on which trout depend for their existence.

At one time these little brooklets and drawings were valued, but little is now done to keep them pure and wholesome. So it has become part of my work each year to see that water has a passage along them to keep clean the bright gravel of their beds. And to be nurseries for trout and trout food, these little streams must have attention. They are quite as important as the main river, in fact, more so, for to-day the water of our main rivers is no longer the pure element that is so necessary for the reproduction of the aquatic creatures which should have existence in them. Each year one has to rely more and more on the tributaries.

For natural regeneration it is essential that we should

have streams that are free from any pollution—places where the small life has a chance to exist until large, or strong, enough to combat the impurities of the main river course. We must have somewhere in which the trout can spawn, and where there is a possibility that the eggs will produce young fish, and where, not only will regeneration take place, but the young fish find a food supply that is satisfying during the delicate stages of their early life.

It is additional work for the river keeper to have to maintain these little waters, but I consider it to be well worth while. Each little brook can produce something of value, and I think such streams should no longer be classed as minors, but be considered as very important assets to the fishery. But it is useless wasting time and energy in cleaning them out if, linked with some habitation, place or roadway somewhere along their course, there is a drain, or drains, from which impurities are likely to be discharged. In such case it would be much better to allow the stream to fill up with vegetation and rubbish which could act as a filter before the water could enter the river.

However, some of them are perfectly pure throughout their course, and these can be prepared so that they entice the spawning trout to use them, and so that they can become a breeding place and sanctuary for many other of the river creatures. Where it is the policy to stock with unfed trout fry, what better nursery is there for little fish than a little river which contains everything that is necessary for their growth and shelter.

Sad though it is, I am very doubtful if there is a river or stream in this country at the present time that is perfectly wholesome and as natural as it should be, and, I fear, even though every precaution may be taken to combat pollution, certain impurities will continue to enter into our waterways and in sufficient quantity to have adverse effect on many of the aquatic creatures.

Now, at various times, I have read comments on the value, or otherwise, of vegetation in a trout fishery. Though it is generally admitted that vegetation is helpful in many ways,

12

I have noticed that one very important and valuable asset has often been disregarded. I would like to draw attention to the great help weed-beds can be in purifying the waters of rivers and streams, for in these days we should make all possible use of anything that is of assistance to cleanse the water.

The fact that weed-beds can help was brought home forcibly to me when a sudden and very heavy storm swept part of the Upper Avon Valley late one June. Thousands of gallons of drainings from several roads and farm tracks entered the river at the head of a long reach of shallows— a surging volume of rainwater which, as it washed down the chalky tracks and lanes, down the sides of the hot, tarred roads from the hills, and into the gutters and drains of the village, had changed to the colour of very dirty milk and was impregnated with filth. As such it was being discharged into the river.

The level of the river was lower than usual at that time of year and I watched in disgust as the filthy muck mingled with the clear water of the hatch pool—watched until, from bank to bank, the water quickly changed colour and then swept with the current down-stream. Such water was unfit for anything to live in. I saw the dorsal fins of several grayling break the surface while hordes of minnows and other small fry crowded to the edges of the river where water, out of reach of the currents, had remained clear. Here and there flies hatched, to struggle and flutter about for a while before floating away, trapped in the surface film.

Fearful for the safety of the insect life and trout farther down the river, and with the intention of watching carefully anything that might happen, I went towards a set of controls a mile down-stream. The Mayfly season had finished, but so far no weeds had been cut in this particular reach. All down it, great masses of flowering ranunculus were spread across the surface from bank to bank—masses which at intervals were so dense that passage of water through them was impeded, and where clear spaces were almost negligible. Into these weed-beds swept the dirty water.

13

I walked slowly to keep pace with the discoloration, but as it passed from weed-bed to weed-bed I could see it was gradually clearing. Halfway down the reach and in a short length clear of weeds, the river bed showed plainly. There I waited. After half an hour the gravel bottom was still visible, though the water had clouded a little. The danger I had feared was passing away.

Retracing my steps and going up-stream quickly, I came upon cloudy conditions. After the first hundred yards I could no longer see the gravel bottom—a farther quarter of a mile and nothing but the tops of the weed-strands was visible, and at the pool above, the volume of filth entering from the ditch had increased. Once more I went down-stream.

Though it was clearing a little as it mixed with the greater body of the clean water in the river, it became increasingly obvious that the dense masses of ranunculus were acting like a great filtration plant. The widespread fronds and tresses were gathering and absorbing the filth from the water acting like great strainers to purify the river and render it once more fit for the aquatic creatures to live in.

In face of this kind of thing it is easy to understand why few rivers at the present time are in the pure, wholesome condition required by Nature to enable regeneration of trout to take place, and why we have to resort to certain artificial methods to assist in maintaining a stock. Of these, the most widely practised is to catch up trout of both sexes when they are in ripe spawning condition. The female is, by gentle pressure, stripped of her eggs and the male is similarly treated for fertilisation, then both are returned to the water unharmed. The eggs are laid down in hatcheries where, if the simple task of fertilisation has been properly carried out, a very high percentage will hatch into alevins and eventually become fry.

So far so good. By the discovery of this method of artificial spawning and incubation, a great gap in trout production was bridged, for even though a water is unfit for the natural hatching of trout eggs, there is this means of providing each year a sufficient number of little trout ready for distribution into the fishery. What is more, it is possible to ensure that these tiny fish can live through one of the most delicate stages of their lives—namely, through what is known as the alevin stage. I feel sure the most vulnerable part of the life of a trout is during the first two months of its existence, and the success of a wild trout fishery has to depend entirely on the available facilities for the production of the right kind of food during these first two months.

As one delves into Nature, so more and more things become apparent. At first one wonders why trout should choose to spawn in winter, and then why the freshly hatched fish should be encumbered by an ungainly yolk sac, but in all Nature there is a reason for everything and in trying to assist her we must find out these reasons and try to follow them through.

First, the spawning. It becomes obvious that during winter the temperature of the water drops to its lowest point of the year—to a point when all animal and vegetable growths are at a standstill. The increased water supply welling from the springs cleans all the foulness of rotted vegetation from the gravel and with the additional aeration this cleanliness is maintained for eight or ten weeks. Trout eggs laid in the gravel during this time will remain clean and well aerated, and are unlikely to be affected by animal or vegetable growths.

So, provided the water supply is pure, and egg laying and fertilisation are successful, long before the river temperature rises sufficiently to awaken the tiny animal life, the eggs hatch into little fish. The low temperature has also kept back vegetable growths which could be harmful to both eggs and alevins. Now, though water conditions have been favourable for incubation, they are not favourable for the food production needed by these small trout. Water is barren of all life suitable for the existence of tiny fish; a period has to be bridged until food supplies can be available, so Nature has provided each little trout with an umbilical sac, from which a sustenance can be absorbed. This sac lasts them from six to eight weeks, and feeds them while the slowly rising temperature of the water gradually gives life and sustenance to the vegetation and animal kingdom which is to provide the initial food for the young trout.

In the south, trout usually spawn during the months of January and February. By March and early April the contents of the umbilical sac are exhausted, animal life and vegetation have sufficiently advanced to provide food for the little fish which now have taken up positions near the edges of the river or stream. Trout fry need a certain food during the first month of feeding and if we are to help them through this precarious stage then we must find out just of what this food consists.

Some say this initial food consists mostly of plankton, but my own studies have proved otherwise on the Avon. March is much too early for movements of plankton to take place, for this happens in the warmer months of April and May. In autopsies I have carried out on little trout during this first month of feeding I have found the food contained in them has been exclusively the larvæ of midges.

My careful study of small wild trout convinces me that they depend entirely on water-borne animal life. They poise themselves in mid-water and just wait for food to come to them. By this it is obvious that the food supply is contained in the bulk of the water—that during March and April, and only at this time, the river is impregnated with the life which forms the sustenance for them, and midge larvæ form the bulk of the life carried by the water during March. I know them well. I know also the tiny insects into which they eventually transform, and I like to think that Nature provided these creatures solely for the food of baby trout, for I cannot discover another reason for their existence.

Migrations of the larvæ take place in March and early April. Many millions of them move from place to place, each and all seeking a situation where they can pupate. This movement is confined to the shallow reaches of river and stream. In April they hatch, and on warm afternoons and evenings the midges return to the water to lay their eggs.

16

Each little female midge, small as it is, produces a large quota of eggs. Often they are busy egg-laying in millions—countless numbers of tiny weaving creatures all dipping to wash off their load of eggs which, when developed into larvæ, will in the spring of the following year produce the main food for tiny trout.

I could wish that the discovery of this essential food was the end of the problem, but it is far from the end. We must go still farther. Much work has still to be done, for, essential as they are to the existence of trout, we must delve into the life cycle of these tiny midges and find out just what environment is most suitable for their well-being and then, perhaps, help to provide a food for them.

It was a doubly sad blow for trout fisheries when the irrigation system was abandoned throughout many parts of our river valleys. Water meadows were helpful in lots of ways and I hope in the future that our farmers will once more value the grass and other good herbage which can be produced by irrigation. However, my interest in water meadows is not so much concerned with the production of first-class cattle food as in the fact that the multitude of waterways which make up an irrigation system all have enormous value in producing insect life and trout.

No wonder it is said that there were more trout and insect life in the days of long ago. Of course there were. Fifty years ago the waters of our valleys covered a surface almost treble that which they do to-day. Our river courses then were just recipients of hundreds of tiny waterways, and of the multitude of creatures which each and all of the waterways produced. I have no need to read of what happened fifty years ago nor to listen to people older than I who knew the meadows at that time, for I well remember my own boyhood. I can recollect the times when I wandered about the meadows in the Avon Valley at the end of April and in early May—times when the water had been shut off from the irrigation system to allow the meadows to drain and to dry, and for the sun to complete the work the water had started.

17

Those grass-fringed runnels, ditches and carriers remain vivid in my memory to-day, for everywhere was a scene of activity as fish and insects hurried to get into the drains and drawings before they became trapped. Everywhere were masses of crawling caddis and snails, every stone or loose turf sheltered a horde of shrimps and small fly larvæ, while thousands of minnows scattered in panic as great trout and eels sped along the carriers to safety in the deep pools at the hatchways, or scrambled with their backs out of water, to the drains, and thence to the main river. When the meadows had drained, every pool, from the tiny bunny-hole pool to the main hatchways controlling the irrigation system (in fact wherever water remained), was full of life, as creatures of all sizes and descriptions vied with one another in finding hiding places in which to escape observation or attack by enemies.

Yet these hordes I saw trapped were but a fraction of the life that had been present in the meadows but a short time previously, for the drains and drawings all linked with the main river course, and the bulk of the insect and fish life had drifted out of the meadows with the draining water. These creatures had taken up new quarters in the river or in the fresh flowing streams which continued in the main drawings throughout the summer months.

Though called drawings, these waterways were the main drains of the meadowland. They were in the lowest part of the valley, for they had been originally the natural river course until man saw fit to create others which carried the water to a level above the surrounding meadows. And being the main drains they were always well looked after and kept clean and tidy by the drowner in charge of the meadows. Their beds were bright and gravelly, and this fact was greatly appreciated by the trout, which during winter, ran up these sidestreams to spawn. It was here that, later, the trout fry found a food supply that was plentiful and sustaining, until such time as they were big enough to drift downstream and live in the main river.

But now, in much of the Upper Avon Valley, and other

valleys too numerous to mention here, the irrigation system is a thing of the past—an almost forgotten art. The waterways of the meadow no longer produce trout and trout food and act as nurseries for the main river, for everywhere is a scene of neglect. The carriers, which in bygone years brought life-giving water to the meadow vegetation, are now dried-up ditches—drains have been trodden flat with the countless hooves of cattle. Bridges, arches and bunny holes have all collapsed, hatches lie rotting by their structures, or are suspended high and dry in their sockets, while the crumbled wingings of wood and brickwork of the hatchways now rest amongst the dried mud and other rubbish which fill places that once were aerated pools of water—the homes of innumerable creatures.

In these circumstances it is inevitable that there are fewer trout and less insect life to-day, for now, in many places, the main river course is the only waterway in the valley, and often this is drained to its lowest possible level so that half its bed is dry during the greater part of the year.

The drowner's job was a vital one in the old days and even to-day, though there are few of them left, old hands at it linger. They are an independent lot and have their own ideas, as I found when I discussed the old times with one of them.

He could recollect the valley and the river of his youth—more than sixty years ago—and I questioned him concerning the flow of water down the river course at that time. His answer, which was not what I had been expecting, is worth quoting in the speech of our valley: " Jest thee look et thay withy bids en thay thir wold pollerds vallin en sprawlin all awver tha placin. Al on em be drinkin ep tha water is vast is a thirsty 'oss."

The words of the old drowner set me thinking, for here is water extraction with a vengeance and a source of loss that I feel sure cannot have occurred to many. I write mostly of conditions as they are at present in the Upper Avon Valley—though I know these differ very little from those farther down-stream, or from other valleys in the south of England.

Much has been written of water extraction and many have deplored the installation of the various pumping stations which here and there take the water we so badly need for our fisheries. Some of these pump at the rate of half a million or more gallons a day and we have statistics to show that this pumping has materially affected the flow of water down the valley. But here is something that is perfectly obvious: we know, almost to a gallon, how much water is extracted during the months of summer and autumn—the seasons when water extraction is most vital to the life of a river.

Seth, the drowner, was able to give me a very clear picture of what happened long before I was born. "Ah," he replied to a question, "thay medders wir tinded thin, in us used ta cut thay withy pollards en thay ashes fir vaggots, spars en vincin poasts. None on em wir lowed ta git like thay be now. All thay withy bids wir planted zince I kin mind, en us used ta keep em cut. Zum on em wir used ta mik idges ta kip in tha ship en tha lambs. Thin thir be thay alder bids. I elped plant zum a thay—varmer wir gwine ta use em fir brumeads, er zummit, bit twer niver done. En thay be zum thirsty varmits—wuzzin withy I low."

On further conversation, of which, being a Wiltshireman myself, I could readily understand, Seth was of the opinion that both withy and alder are thirsty "varmints"—that they drank more water each season than trees such as beech or oak, even though such trees were "tin times as big." "Thee ave a look at thir roots en zee all the vibres thay got pushin out into the ground, then thease know what I da mane. All thay pinky little shoots wants a drink s'now."

Now, as a rough estimate, I would say there are more than 1,000,000 withy and alder trees in six miles of the valley I know so well. Seth estimated that each tree, throughout spring and summer, needed a gallon or more of water on which to thrive each day. So from this it meant that approximately 1,000,000 gallons of water is being extracted each day from the low-lying ground—from the supply which otherwise would bubble up in springs to feed the river. Just think of it—1,000,000 gallons of water, though not much, could help the flow down the river course each day.

As old Seth wisely remarked, "T'wouldn't matter a cuss if thay drunk ep tha water during tha winter, but thin thay doan't, s'now. Thay be idle critters thin, whin tha could help thee git rid a zum ev tha vlood waters." He is right, of course. Trees need very little moisture when they are bare of leaves.

When these things are explained one can begin to realise just why things have changed. After all, it is very doubtful if there were many trees of any kind in the valleys of a hundred, or more years ago, for where can you find any trace of their origin. Doubtless there was more water in the low-lying ground and more flow down the river course, and it is equally obvious that as the trees grew larger and increased in number, so they needed more water each day.

I have not written this because I would like to see all withy trees and alders cleared from the valleys, far from it. Many of them, old and decrepit as they are, have become very familiar, and I think add to the charm of the riverside. A river valley bare of trees would be a sorry sight indeed, but before laying all the blame for water extraction on to the pumping station as a reason for our slowly dying rivers, it is as well to look at Seth's point of view.

3   THE TROUT

Whatever the water, the angler's prime interest is in the fish it contains. Often, even with practice, it is far from easy to find them. With the exception of winter, when trout are preparing to spawn, or actually spawning, one would naturally think that they must feed at least once or twice each day or night, but this theory can frequently be proved wrong. Time and again I have caught trout and examined their stomachs to find them empty of food, and it has been quite obvious to me that no food has been taken during the past day or so. Beyond this it is impossible to say, but it is more than likely that the fish have taken little or no food for a week, or even longer, without any apparent ill effects.

Though one can never be sure that trout sleep, I think there is ample evidence to be obtained that, if not actually sleeping, the fish do lose interest in everything around them, and for long periods will retire to some secluded position where no creature, in the natural course of events, is likely to interfere with them. There are few days in a fishing season when all the trout in a river are feeding at the same time, and I feel sure there is never a time when all trout are rising to surface food. The sport of dry-fly and/or nymph fishing mostly has to depend on the odd fish here and there which, for some reason or another, have chosen to feed near the surface. But there are occasions when even the customary few are absent, when you can walk miles and miles along the banks of a fishery without seeing a single trout that would be worth the sport of catching.

Sometimes days, even weeks, will pass without any change and you begin to be assailed with doubts as to whether there are indeed any trout in the water. Questions pass through your mind. Can anyone have poached the fish, or anything happened to destroy them? Have they all migrated to water up-stream or down? Have a few big pike established themselves? Then one day all your fears are set at rest, for there, appearing as if by magic, are trout of all sizes and condition, and each on the feed.

It may seem that conditions are no different from the previous day, or those of a week earlier. Hatches of fly have not increased to any extent and sub-aqueous life is no more active, yet something has happened to bring these fish into positions where once more they can be seen from the river banks. I think trout are like the pike and the eel—that a period of rich feeding is sufficient to last them for a long time. If you make a thorough search of the river during periods when, in viewing from the banks, the absence of trout is most noticeable, you can discover the fish. As you search, so you are surprised, for you will find the trout tucked away in all kinds of places and to all appearances in a state of coma. Most of them will be completely hidden beneath the banks under tree-roots, in weed-beds, or perhaps a score or more will be found hiding beneath a bridge where the supports have been undermined by the water. Furthermore, there may be debris lying on the river bed—a tree branch covered with weeds, a boulder, a tin can or sheet of iron, each with a hollow beneath it sufficient to form a hide. Some, however, will rely on their power of camouflage, and lie flat on the bottom with scarcely any cover above them, and so well do they merge with their surroundings that when wading one may inadvertently touch them.

Have you ever disturbed a big trout day after day and watched its reaction to what it thinks is danger? Often it can be enlightening, for you will have noticed that the fish invariably bolts to the same place and there takes cover. Trout are naturally shy. They depend to a great extent on their powers of camouflage and on their ability to find and take shelter in some impregnable position. Throughout the years they have inherited an habitual shyness of man, and in many rivers and streams trout must now consider man to be enemy number one for, in some places, the majority of their natural enemies are controlled, if not entirely eliminated.

It is a mystery to me how trout are able to discriminate between men and cattle, or dogs. Yet they can and do, for many times I have watched cows and horses walk past a feeding trout without disturbing him, yet that same fish has bolted for cover the moment I came into his sight. But the point I wish to make is that it is quite obvious that if large trout are to feed contentedly in a river, then, nearby must be facilities for hiding places. It is well known that the trout feeding in mid-stream is always more shy—more difficult to approach, or to catch—than his neighbour feeding close under a bank. To a certain extent, Nature provides facilities for hiding places. Banks are hollowed, pools created and weed-beds are given life. Here and there big stones and logs lie on the river bed, while tangled masses of rushes fringe the edges of the water. All of these, and others, will provide a hiding place for a trout—a place where he feels secure. Perhaps more than one will choose the same site as a sanctuary and live in contentment on the food that is available nearby. These are their homes, the places to which they can retire in time of danger or when in need of rest and seclusion. When movement of insect life prompts them to come out and feed, they do so, safe in the knowledge that nearby is a holt in which they can quickly hide.

A given reach of water may hold several hundreds of large trout. But should all their homes be destroyed, as is so

often the case where dry-fly fishing is the practice and rods
want clear, open water in which to fish, the result is that
either the fish remain near the area of their previous home
and bolt wildly in any direction the moment they sense or
see danger, or they congregate into one of the deeper parts
of the river or stream, there to become easy prey for a big
pike, or of the first otter which discovers them. Some may
rise to fly during daylight, but the majority are afraid. They
are like wanderers in a strange land and are continually on
the look-out for danger.

This is the importance of weed-beds and other cover
in a trout stream. When you have a number of big fish in a
reach, it is much better to have them spread out as thinly
as possible. If Nature has failed to provide the necessary
conditions for their peace of mind, then she can be assisted
in many different ways. One thing is quite certain. If trout
are to rise and provide sport during the daylight hours they
must have a place for retirement. It pays to leave some of
the holts and hides even though they may be obstructions,
otherwise, being the shy creatures that they are, the feeding
times of big trout will coincide with the solitude of early
morning, and with the cover afforded to them by the
darkening sky of late evening.

During hot summer days many fishermen must have
heard that queer sucking noise which at regular intervals
comes from a weed-bed or from somewhere under a bank.
In the South it is heard mostly where long tresses of ranunculus are spread in a dense mass over the surface. Often the
sounds go on and on without, at a distance, the slightest
trace of their origin. Years ago I had been told that these
noises were made by eels as they fed in the weed-beds and
one day, having nothing better to do, and thinking I might
get a few eels, I went out to see if what I had heard was
indeed true. Taking with me a five-pronged eel spear, I
walked to a shallow where, during the last week, the noises
of sucking had been very frequent. Nor had I long to wait.

The sound came from several different weed-beds so,
stealthily wading out to the nearest, I waited for a sign of

movement. My approach caused no alarm and beneath the weed strands I could just sense a shape and then the vegetation was thrust upwards within striking distance of my spear. In this brief moment the nose of a fish appeared, but the nose of an eel is not unlike that of a trout. Down went the spear. It is true I got two eels that afternoon, but before I fully realised just what was making the majority of the sucking noises, my spear had transfixed half a dozen fine trout. I had learned another lesson. It pays to find out for yourself before trusting to the information of others.

Since then I have taken every opportunity to study trout when they are feeding in this manner. It has been interesting. With close study there can be no doubt that it is mostly trout which make the sounds, for I have watched them time and again, from distances of a few feet and indeed some trout have continued to feed when within inches of my waders, and when I could quite easily have touched them with my hands. It would seem that they feel perfectly secure amongst the strands of weed, and if you remain perfectly still, they are not in the least alarmed.

But though close at times, it is not easy to see a trout moving about beneath the weed tresses, even if occasionally it is possible to follow his course, as he touches strands of weed and makes them quiver at the surface. This disturbance is always very slight. The trout is hunting, and the catching of his prey depends on him moving warily. Slowly he swims about beneath the vegetation with eyes continously searching the surface. He knows that just beneath the water level mature nymphs are clinging to the fronds and stems, there waiting for the cool of approaching night before hatching into flies. And there might be several different species and genera, but the majority will be Spurwings and Blue-winged Olives. As each is sighted so the trout moves towards it.

No hurried movement this—he knows his prey is within easy reach—and then comes a gentle opening and closing of jaws on insect and weed-stem. For just a moment, the nose of the fish appears above the water with the weeds held in tightly closed jaws and to the fisherman on the bank comes the noise of a distinct suck as the insect is parted from its hold and the strand of weed is released.

Sometimes it pays to watch the down-stream end of a bed of weeds, or any area where trout can be heard feeding as I have described. But one must be very patient and be ready to cast an artificial accurately the moment an opportunity is presented. Usually it is good-sized trout which feed in this manner. It is useless to try to interest the fish while he is beneath the weeds, but occasionally his search will bring him to the tail end of the weed-bed, where he may enter open water before turning up-stream to work through the bed once more. His appearance will be very brief, but he is searching for nymphs—expecting to see them—and a good representation placed within his sight is taken without suspicion.

One might perhaps wait for half an hour, or more, while the sounds of sucking come from all parts of the weed-bed, but on those hot summer days when one is whiling away the hours until the time of the evening rise, there is just a chance that patience and keen eyes will be rewarded.

I suppose it is just human nature to gloat over the capture of a big fish, but though there is certainly a great degree of skill needed to play the victim out until it can be landed, the actual hooking of a monster is just a matter of being lucky enough to be on the spot when he is feeding. I have found many times that a big fish is much easier to deceive than a smaller one.

A trout of exceptional size in any water is one that has probably spent most of its time in some deep secluded spot; his eyesight is often failing, and his sense of self-preservation has been dimmed by the fact that he has lorded it over the other occupants of the river so long. And unless there is something which he urgently needs, or which he calculates will provide him with a meal in a short time, he rarely comes to the surface to feed, but prefers to live on food he finds near the river bed—food which is large and satisfying.

Such a fish might be anything up to twenty years old,

27

and possibly there are times when he is in need of a little stimulant, or insect life with a medicinal property, and at such times will rise to take tiny insects such as black midges, caenis or greenfly. But, in any case, whether the attraction is big flies such as Mayflies, Sedges and the like, or the very tiny insects on which he occasionally feeds, when a big trout comes to the surface for food, he really means it. Often he chases away a smaller fish from a choice position and then takes insects from the surface in a carefree manner, as though he is anxious to obtain what he wants as quickly as possible so that he can return to seclusion. One seldom finds a big trout rising for any length of time. I speak of the big fish as a he, for in ninety cases out of a hundred the monsters of a river are males. A big trout may rise just for one short period during a season. If he is lucky, he rises at a time when anglers are absent; if not, it is the fisherman who is lucky.

Possibly, when younger, the fish had some experience of artificial flies; perhaps he was hooked and lost. No doubt he would retain memories of such an experience for some time afterwards, but I think it is extremely unlikely that such a fish could keep a vivid memory for one, or more, years. So when an artificial floats towards him or a nymph passes near his nose, he has much less suspicion than fish of smaller size, fish in the prime of life, which are regularly being cast over. The big trout often takes the first offer without hesitation. Many of the big fish which are caught have been surprises. The surprise comes to the angler when the hook goes home, for often the cast is made to a rise form in a position where a much smaller fish has been known to feed. And without question some quick thinking is then necessary to handle the situation successfully, for a big trout knows full well in which direction his home is, and he makes little hesitation in going to it.

Sometimes I think that there is confusion over what is meant by the term "cannibal trout." It would seem that many writers like to imagine these to be the ogres of a river, and that a trout has only to reach a certain size before he habitually becomes a cannibal. Many a time, when I have been looking into some deep hatch-pool with fishermen, they have remarked that a few old cannibals must live in these dark, secluded places, and from their tone of voice I can tell they think one or two monster fish are lurking there, just waiting for one of their unsuspecting little brethren to come along to provide a meal. Often there are big trout in a deep hatch-pool, or, for that matter, any other deep pool, and without doubt they are cannibals—if by cannibalism it is meant that they eat their own kind, as well as other species of fish. But what I would like to know is why, in the minds of some people, a trout has to be a big one before it can be termed a cannibal.

All trout are cannibals. From very early in life they have no tender feelings towards any of their own kind that are smaller than they. A trout is a cannibal from the time his stomach has grown large enough to accommodate a smaller fish, and if one is to kill them just because they feed on their own kind, then why concentrate the hate on the big ones.

A trout of, say, 4 lb. making occasional meals of $\frac{1}{2}$- and $\frac{3}{4}$-pounders is doing no more damage in a water (from the point of stock reducing) than the fish he has eaten, for frequently I have found fry in the stomachs of trout which have weighed less than 1 lb.

Many times I have been asked if cannibal trout are good to eat. Of course they are good to eat—in fact, the big fish should eat better than the smaller ones. I think people get a bit confused with their cannibals. A trout feeding entirely on lesser fish is usually in excellent condition. Like a pike, he is healthy and strong. Later, as old age creeps on, he may be a horrible looking brute, with big head and eyes, and with a lower jaw that has a hook like a salmon kelt. In the last stage of his life he is a brute horrible to look upon, and one which gives little sport in the catching. And it is these brutish-looking trout which are so often described in fishing books and articles as cannibals. I think scavengers would be a name more fitting to them, for they live on the dead as well as on the living, and as such they should be

29

destroyed whether of a length of 10 inches or of 30 inches.

My experience has indicated that another horrible disfigurement of trout, blackness, is independent of age and can be traced to two causes: injury and the inability of the fish to shed their ova, thus breeding thread worms. Injuries can be caused by all the natural predators of a trout stream—by herons, pike, otters, kingfishers, etc., or by wounds from other causes. Twice in a season I have had the opportunity to see good coloured trout gradually turn black after being injured by pike. I now feel sure that the pigmentation system of a trout can be completely destroyed by laceration of the nerve centre which, apparently, is on the lateral line.

Both the trout I watched were fish of just over 1 lb. in weight when I first noticed their wounds. They were in excellent condition. One was in a place where I could make almost daily observations, as it was one of a number that were fed by my father every morning. It was in early May when I first noticed that this trout had been slashed by a pike. The fish was quite active and took food with the others. Along each side, from the tail to the dorsal fin, I could see the open wounds where the teeth of the pike had slipped as the trout struggled free. But about a week later I saw that discoloration was taking place around the area of the teeth marks, and then day after day a gradual change took place until the whole fish was as black as a bowler hat. By the middle of June a marked difference in condition was also apparent, and though the fish swam about with others after food it did not take any.

But now I could no longer see the marks of the pike's teeth. It appeared as if the wounds had healed perfectly and that scales had formed once more. As the weeks passed so the fish became less active and took to lying flat on the bottom in a sheltered place at one side of the pool. This fish had interested me and I had watched closely in the hope that it might regain its colour and condition, but by the end of July I could no longer stand the sight of the now ugly brute, so I caught it with a landing-net.

The fish made no effort to escape as I put the net ring

down over it and it made no struggle on the bank. A sharp rap on the head quickly ended its miserable existence, and I then made an examination. So completely had the wounds healed that, had I not known what to look for, the teeth marks might have escaped my notice. The fish was no less black in the air than it had been in the water, and it was blind in both eyes. In about three months this trout, by injury from the teeth of a pike, had been reduced from a healthy, good-coloured fish to a blind, black, emaciated creature, whose death was but a matter of a few more days or weeks. I have not the space here to write of more than one individual instance, but I have seen others that have been equally convincing. I have found that a thorough examination, which includes skinning the fish, will prove in nearly every case that black trout at some time or another have been lucky to escape death. Perhaps it would be better if they died instead of escaping, for I hate to see ugly specimens in a river.

I would suggest that conclusive evidence could be obtained regarding the cause of black trout if all fishery owners, keepers and fishermen made a study of the fish they kill. Often there is nothing outwardly to show that a black trout has been injured, but scars will show plainly on the inside of the skin. If all information was pooled and published I feel sure it would do much to clear up the controversy which for years the subject of black trout has fostered.

All living things must come to an end, for this is the law of Nature. Always the old must make way for the young. But I prefer to think of all creatures in the prime of their lives—to be able to see them enjoying their existence and, in the case of trout, to be able to catch them, or see them caught, when they are in the peak of condition. Each year many thousands of trout are killed in the interests of sport, yet there are others, many others, which die a natural death. These are the fish which despite the danger of predators— pike, otters, herons and others—and the deceit of mankind have lived through their allotted span. They must die as

Nature ordains, but I can find it in my heart to wish they could pass from the world in a manner more fitting to a sporting fish.

The size and weight to which a trout can grow is determined mostly by the available food supply and in every water there is a limit. My experience leads me to think the majority of trout attain this limit in size in about six years, and though they may live to be ten years old or more, they do not, after six years, add to their weight or increase their length. In some waters, this limit may not be more than a few ounces—$\frac{1}{2}$ lb. at most—in others 1 lb. will be the maximum. Some rivers and lakes may produce sufficient food for the limit to be $1\frac{1}{2}$ lb. or even 2 lb., while in exceptional waters the peak of condition will be nearer 3 lb.

But in all these waters it is hopeless to expect trout to exceed their limit. It is true there are a few exceptional fish in every water—fish that are double, treble, or even ten times the weight one can expect to be produced. But just because there are these exceptions is no reason to think that all the fish will live to be a similar weight before they die. One could leave a water unfished for twenty years in the hope that then big trout would be plentiful, but such hope would prove false. One would find (in a water previously producing plenty of trout up to $1\frac{1}{2}$ lb.) that $1\frac{1}{2}$ lb. would still be the weight of the majority of the big fish after the interval of twenty years.

Nothing would be gained—in fact, there would be a great loss, for each year during the twenty, some fish would die a natural death. These would not be giants, but just fish of $1\frac{1}{2}$ lb.—fish which had reached that peak of condition the food in the water is capable of allowing them to attain.

The natural end of a trout is starvation—starvation brought about by blindness. Death comes slowly but very surely and the last stage of a trout's life may extend for two or three years. First, he reaches the pink of condition. He grows quickly to three years old then, with few exceptions, he spawns annually. Rich summer feeding is needed to make up for loss of feeding during winter and for the

32

strength he expends in spawning, but during the summers following the first three spawnings he is able to mend his condition and add a little to length and to girth. But at six years old he finds the balance—the rich feeding of spring and summer does no more than compensate for the loss of condition through spawning. And so it continues.

He may live on for several years—losing weight in winter, gaining it again by the following autumn—but slowly, though surely, he finds it more difficult to maintain the balance. As he ages, so his eyesight fails. Younger fish are more active in taking the food supply. Spawning saps his strength, and no longer can he live where the maximum food supply is to be found. A year or two may pass and though he goes to the spawning grounds, he no longer takes an active part. Summer feeding becomes more and more difficult and soon he is forced to feed by sense of smell. Gradually he gets thinner, and now weighs no more than he did at three years old. He seeks the backwaters and quiet places about the river—places where he need no longer use his fins and strength to maintain a position.

Weeks and months pass along and sightless eyes are in a head that is out of all proportion to the body, and soon all desire to feed is gone. With the instinct of the wild creature he seeks a place to die and with his last remaining strength pushes his way into some cover, on, or near, the river bed. Little flesh is now left upon his bones and no air contained within his body. There he lies concealed from the view of his fellow creatures and then, with a few gasping movements of the gills, his life is over.

So ends the life of a trout, but Nature cannot allow his emaciated body to pollute the stream. To the scavengers of the water comes the scent of death. Crayfish and shrimps gather to do their work, and soon the bones of a once noble fish are scattered over the river bed.

4   REGENERATION

Even though I have seen the same thing many times, the sight once more of scores of trout spawning on the shallows gives me the greatest of pleasure. Many of these fish are of a sporting size, and amongst them are trout which would grace any fisherman's basket during the fishing season. It does me good to see the waves and turmoil, the flashes as the light reflects from sides and bellies, and dorsal and tail fins showing above the surface, for I can picture these fish later on, waiting in some run or eddy, under a branch or jutting bank, sucking in the luckless flies which drift to them with the currents.

When the spawning grounds are in perfect order the river is under control and gravel showing bright and clean under the clear, rippling, aerated water. Here and there emerald green fronds are spreading from early ranunculus, whose bunches help to break up the bottom currents and form them into fast runs. Then I have no difficulty in seeing the trout busily occupied in making their redds. The general spawning season for the Upper Avon is (unless river conditions are adverse) usually the six weeks from about 1st January until the middle of February, but in 1934 and in 1948 I recorded the first pair spawning during 19th December and the last pair during 13th March the following year—a spawning season which extended in both cases for nearly three calendar months.

These early and late fish are what I term out of season spawners, and the reason is not far to seek. Most of them are small fish, spawning for the first time at the age of three years. They are fish which have been introduced as fry—fry which hatched from eggs imported from other fisheries about the country where the spawning season has been either early or late. And for a season or two these fish reach maturity at a time similar to their parents. After a while they adapt themselves to the river conditions in which they live, and then spawn with the natives. I base this theory on the fact that large trout, fish of five years old or more, are seldom early or late if river conditions are normal. I also know that the exceptionally long seasons of 1934–35 and 1948–49 coincided with the years 1931 and 1945, when the fishery imported eggs from other hatcheries, many of which hatched and successfully reached maturity.

Each reach of the river has its suitable spawning ground or grounds. Whether these are in the main river or in side-streams they are places where conditions are favourable for regeneration. Year after year the trout congregate at the same place and though they pair off in late October or early November they will remain together and (unless interference occurs from other sources) will spawn together. They seem to have little objection to spawning in a communal setting.

It would appear that the knowledge of these locations is instinctive, and that each pair of trout individually find their way to places most suitable for them. In the case of the smaller fish which have come from parents living in another part of the country there can be no question of inherited knowledge, and as these little trout often arrive first on the redds it is quite reasonable to assume that they make their own choice independently rather than by following others. It may well be that the trout inhabiting a river recognise certain places as being suitable for spawning and remember them. Sometimes the paired fish may travel a mile or more up-stream to spawn and the female leads the way. The male just follows.

Though much evidence has been published that natural spawning of trout can be a very efficient affair, I hardly think that experiments and research carried out in other countries should be accepted as entirely convincing. One cannot compare the virgin waters of, say, New Zealand, with rivers and streams in the British Isles.

It would be most interesting to see just what research into the subject in Britain would reveal. Unfortunately, the majority of our trout waters are no longer of the purity necessary for the regeneration of aquatic creatures. I cannot accept any statement that natural trout spawning is universally efficient. I have had far too much proof that it is the reverse.

It is quite reasonable to assume that reproduction of trout in the waters of our country years ago was just as successful as it is in New Zealand at the present time. But conditions have altered considerably. Successful fertilisation and hatching of trout eggs must depend on everything being favourable and natural. It is common knowledge that most of our waterways have been tainted and rendered impure through pollution. Most of them have been affected to such an extent that the cycle of trout life has failed. We have been forced to compromise—to try by artificial means to make up for our abuse of Nature. But it is a losing battle.

Natural trout hatching can be successful. This I have proved when examining redds in spring-fed sidestreams—in streams that are perfectly natural and pure. The percentage of eggs hatching in these little brooks is very high, and quite sufficient to maintain the existence of the species, in quantity, if a food supply is available. Unfortunately, these pure streams are few; the majority of trout are forced to spawn in the main river course where conditions are very different and there mortality occurs at an alarming rate. I have dug up and examined redds where not one egg in a thousand was likely to hatch and where the eggs have been killed during their stage of incubation. I have proved that a large number are fertilised and that it is the conditions they have to experience during their sixty to eighty days in the gravel that brings the mortality. The successful hatching of trout eggs has to depend to a great extent on the oxygen content and on the clarity of the water. Should pollution be present oxygen is absorbed, and should the water be continually dirty the eggs become so coated with mud and impurities that even if there is aeration it is of no benefit.

If conditions in the rivers could be made to be exactly as those of the few spring streams I know, then there would be no need to hatch trout eggs artificially, or to be continually stocking waters with big trout, as for years we have been obliged to do to obtain good fishing. Instead of trying to take over the work of Nature, we should endeavour to restore and provide conditions whereby everything could take place naturally. Man can assist, but can never command.

Though we know that trout eggs will hatch in static water, it is not a natural state of affairs, for trout always choose to lay their eggs in water that is far from being static, and often in places where they expose themselves to danger from various sources. In many cases, in their endeavour to deposit their ova in well-aerated positions, they will spawn in water barely deep enough to cover their backs. However, Nature has reasons for prompting trout to spawn in the well-aerated water of rivers and streams, and I have found, during many years of artificial hatching experiments, that if one is to get the best possible results, then it is necessary to follow nature as closely as possible, and lay down the eggs in water that has a good oxygen content. It appears that if trout eggs are to hatch successfully, a certain amount of oxygen must be available in the water, which can, throughout the period of incubation, be absorbed through the shell of the egg.

It is generally assumed that when trout, in the natural state, dig up the gravel bed, it is in their endeavour to cover and so protect, their eggs from marauders. This may possibly be true, but I am inclined to think other main reasons of this gravel digging are: (1) to ensure a safe anchorage for the eggs, and (2) to create a system of increased aeration. The heaped gravel of a trout redd acts as a baffle

37

which throws the current downwards, around and through the heap, in such a manner that the eggs continually receive the oxygen they require.*

Though eggs may suffer serious mortality through lack of aeration, too much cannot occur, for even though the bubbling water gives a continual movement to the eggs, it has no detrimental effect. I know many people are of the opinion that any disturbance of trout eggs is disastrous during the first three weeks or so of their incubation, but I have repeatedly found that provided the eggs have been properly fertilised and are laid down in suitable surroundings, movement is an advantage rather than the reverse. The number of eggs laid down in a given area should be determined by the aeration facilities. I now hatch about 3000 eggs to the cubic foot of water and get about 98 per cent. results, but in my earlier efforts, in semi-static water, I found 1000 to be too great a number, as often I lost at least 25 per cent. after the embryo had formed sufficiently to prove the eggs were fertile.

Yet this concern for their eggs is only one of the reasons why trout spawn in shallow, aerated water, for though it would appear that after spawning has finished the trout just forget their regenerating activities, nevertheless, in spawning in the preferred locations, they have made certain provision for the well-being of such progeny as should result from their quota of eggs. For the clean gravel beneath the rippling shallow water ensures a safe refuge for the freshly hatched alevins and when they reach the fry stage and are in need of food, such food will be abundant in the surroundings, and be easy to obtain.

So when Nature asserts herself and prompts the mature trout to seek a place for regeneration, it is to the bright gravel beds of the aerated shallows that they make their way. There, with the instinct passed through generations, they know that their eggs should hatch, and when spawning in concluded they can retire content with the knowledge that should any of their

---

Sawyer later discovered that it was vital for the eggs to be oxygenated by upward currents of water from the river bed. See *Frank Sawyer, Man of the Riverside* (Allen & Unwin) P.158.

progeny survive security and food will be available.

A hatchery to deal with trout eggs taken from the wild-bred stock is a great advantage on any fishery, for one has the chance to plant out the trout fry at a time when they are beginning to look for food. If I lay down some 50,000 trout eggs in a year, and these are taken during a period of about thirty days the little trout will come on to feed in rotation, and I shall have about a month in which to plant them into the river and sidestreams.

My hatchery is fed by spring water, and this keeps a temperature of around 45° F. The eggs take approximately sixty days to hatch. I then keep them as alevins for about a month, and then they are ready to go into the water as unfed fry. So, though I have some 50,000 unfed fry for distribution, I am able to plant them out without haste—a few one day, a few the next, and so on. The alevins live on their yolk sacs for about five weeks, and then they require food. If this food is unobtainable they die. I make a point of planting the little fish into their future homes a day or two before food is necessary, in fact, while they still have a short food supply left in their yolk sacs. A good guide to when they are ready for food is to watch their movements in the hatchery. A few of each batch are always more forward than the bulk, and these may be seen to poise themselves in the water—to support themselves by their fins and to occupy positions in mid-water and near the surface.

The success of fry stocking depends to a great extent on where and how these little fish are introduced to the water where they have to live. Throughout the years I have done fry stocking, I have not yet found anything better as a means for distribution than a three gallon watering-can, and I can thoroughly recommend it. I use perforated zinc trays for incubators, and these are about 11 in. by 9 in., and 4 in. deep, in which approximately 2000 eggs hatch and reach the fry stage. I find the fry come to no harm if the tray and contents are quickly lifted from the hatching trough, and put into a wide-topped bucket of water. The fry can then be tipped into the bucket. Transference from bucket

39

to watering-can is a simple matter, and, provided the can is first quarter filled with water, the fry come to no harm if poured from bucket to can.

One might have to travel a considerable distance along river bank or sidestream looking for, and pouring the little fish into, places where a food supply can be obtained. As a few fish and some water are poured from the can, so more water can be added from the river, so that at no time are the fry likely to die through lack of aeration in the container. As the little trout still have two or three days' food supply left in their yolk sacs they have a chance to settle in their new quarters and an opportunity to get thoroughly accustomed to any change in water temperature and eventually to find a feeding place.

The question of the value of stocking with unfed trout fry has been one that has caused much controversy, but I am sure that if stocking is carried out in a proper manner, and provided that the water produces some naturally bred fry, a fine stock of sporting wild fish can be reared at little expense, without fear of importing disease. Though eyed ova have been imported occasionally from other parts of the country to introduce fresh blood, the majority of the trout killed annually in my water have resulted from eggs taken from wild bred fish of the river, artificially hatched.

Before the war the total annual bag of takeable trout often exceeded 1500; and on one occasion a tally of nearly 1900 fish, averaging 18 oz., was recorded. The Upper Avon trout usually require four years to attain 1 lb. in weight, and the non-stocking period (1943–45) resulted in a big falling-off of the annual tally in the years 1947 and 1948. In 1947 only about 500 sizeable trout were taken and in 1948 some 650 were recorded. The 1949 season again showed the results of fry stocking, when a total of nearly 1300 trout were taken, and it was soon certain that the river again held a good stock of takeable trout.

These figures go to prove that fry stocking can be successful, but it is useless to expect trout fry to flourish in waters where there is no natural regeneration. A water might be capable of producing food of a suitable size and value for yearlings, and fish of larger size, but for fry this is no good; if one tried to feed a baby with beefsteak and onions the child would refuse it and would eventually die of starvation. A human baby needs special food: a baby trout, likewise, requires something that is suitable. To achieve success in fry stocking, therefore, one must first study natural conditions, and if these are favourable then one must plan accordingly. Trout may spawn well in a water, and yet there may be little evidence of regeneration, but this is no reason to assume the water to be unsuitable for fry. Many things can happen to destroy a season's egg-laying, and if the eggs do not hatch there can be no alevins or fry. The water of the Upper Avon is not good for natural trout egg incubation, and there is always a heavy mortality. Artificial hatching of eggs taken from the wild trout is a means by which the river can be kept well stocked, regardless of natural mortality, for the water does produce food suitable for fry in great quantities.

It is not difficult to rear trout to the unfed-fry stage, but considerable study is necessary to decide wisely where to put them in the river or adjacent sidestreams, so that they get every opportunity to thrive. One would not dream of turning out a herd of cattle where there is only sufficient food for a dozen; there must be food available for the number of little trout put into a water, and fish must be placed in positions where they stand the best chance of getting it. For trout are not like birds or mammals: they have no parents to forage for them, to bring them food until they are capable of finding it themselves; they have to start straight away to look after their own lives. They live individually.

Trout by nature depend entirely on living creatures for food; in a river they never hunt about for a meal. When they are ready to feed they move to the edges of the stream, there to wait individually and indefinitely in some secluded little spot, until the food is water-borne to them. The currents of the water act as their parents—parents provided by Nature— to bring them what food she can create. Some days the larder is plenteous, on others there is a scarcity. But regardless of famine or glut the little trout wait patiently. If insufficient food comes to them to sustain life they die. Along the river edges, especially where the water is shallow and the bed gravelly, these little holts or waiting places are legion, tiny little pockets, eddies and backwaters, all within reach of the current which brings the food. It is in these little places that the unfed trout fry should be *placed*, not tipped in by the thousand, or even by the score or dozen, but just one or two at each location, sufficient for the food supply.

It may take a day to plant out one thousand fry; they may extend over two miles of water, but if they are put in places where they can feed the job is worth while and a needless waste of life is saved, for if these small trout had to find their own feeding sites many would be lost or would die in the attempt. It is useless putting them in mid-stream: they cannot stand the strong currents, and cannot feed. It is waste of time (and of life) to put them in deep water, for there no food of a suitable nature is likely to be found.

In a day a thousand hungry fry need at least five thousand insects to satisfy their needs. Five thousand insects, however tiny, is a very large number for any small area of water to produce day after day, especially when these insects have to be water-borne to the mouths of the little fish. Again, enemies of trout fry are numerous. A fowl would eat two thousand grains of corn in a very short time if they were available in one heap, but if these same two thousand grains were scattered over two acres of land, that fowl would take months to find them all: in fact, many of the grains would be growing, and beyond the reach of the fowl as a food supply, before they were discovered. It is the same with trout fry. If they are tipped into a river regardless of quantities, a hundred or more may try to shelter under one big stone, or in some small growth of vegetation. In many a case beneath the stone is a bullhead, or in the weeds is an eel, or large-sized trout; each would welcome a meal so provided. Not only fish but large-sized insect life will attack and kill tiny trout when they get the chance, besides which there are feathered predators to be considered.

But if the little trout are distributed widely each of them has a chance to survive. A foraging enemy finds it extremely difficult to catch an individual trout fry, and even if he succeeds the loss is only one, instead of hundreds. Why go to the trouble of creating life with one hand, so to speak, if with the other hand that life is to be wasted?

For several years I have been experimenting in the feeding of trout fry with natural food, and the results obtained have proved beyond doubt to me that not only can fry be reared by this means but that their growth is increased enormously. The huge loss that is usually experienced when artificially feeding fry during the first month is reduced to a minimum. The proportion of loss under natural conditions had been assumed to be little less. It appeared to me that the point of importance was to provide a rearing place where the minute larvæ, on which the fry feed, would flourish and be abundant. The surroundings must be natural for the fry and natural for their food.

As a result of experiment I found the chief food of trout fry in my water, during the first two weeks of the feeding stage, to consist exclusively of the minute larvæ of one of the *Chironomidæ* midges. These appear to thrive in well-aerated, shallow water of low temperature, contrary to the usual habits of many of this genus. The creature is about one-tenth of an inch long in the spring and may be called a "free swimmer," as it is often to be found swimming in midwater; its method of swimming is by a series of contractions of the body, very similar to that of the gnat larvæ which is found in water butts and stagnant water. This animal feeds on some of the lower forms of animal life, a form of algæ, which is bred in well-aerated water and may be found as a slime attached to stones and some water weeds, the water celery and water cress in particular. It spends most of its time on the stones and weeds in a network of tunnels through the algæ, but at intervals during warm days it will swim into free water and is then taken by the fry.

After the first two weeks in the fry stage, the food of trout fry consists of the minute larvæ of such of the *Ephemeroptera* as have swimming nymphs. At this stage the fry are able to hold themselves in the fast-running water and take these larvæ as they are swimming through the water from, and to, weeds and stones. The movements of these insects are very fast, and in consequence the trout fry have to be quick to catch them; their movements often being too quick to follow with the eye.

The fins of trout fry, especially the pectorals, are very large compared to the size of the fish, if the fins of say, a two-pounder, are taken for comparison. I suggest this development has been arranged by nature so as to give it the speed and power to hold itself in the fast water and successfully to catch its prey.

As a result of these and other natural observations, it appeared to me that if I constructed a stew on natural lines and fed natural food to the fry, good results might be obtained. I constructed a small stew in the open made entirely of concrete, 5 ft. wide by 18 ft. long and 1 ft. 6 in. deep, supplied with water direct from the river, through a 6-in. pipe, the water having a 1-ft. fall into the stew to give good aeration. The outlet was made 1 ft. wide with a sluice, the water being maintained at a depth of 9 in. Fine-mesh screens were fitted to the inlet and outlet. I then covered the bottom of the stew with gravel from a nearby stream and silt from the river, to a depth of 2 in. I allowed the gravel to dry for several days, until I was reasonably sure all animal and fish life that might have been in it was dead, and then allowed the water to flow through the stew for several months. Bearing in mind the natural habits of trout and trout's food, I constructed a series of runs by planting bunches of celery and water cress and placed large stones to break, and make, currents of water. These, after a while became covered with algæ, which grow well in the water, considerably more being found at the inlet than at the outlet end of the stew.

The following spring I found the larvæ of the midge

before mentioned present in great quantities. These were feeding on the algæ on the stones and weeds. I came to the conclusion that these are the larvæ of a midge I had seen in great quantities laying its eggs by dipping on the surface of the stew the previous summer. The first fry, about 300 of them went into the stew on the 15th April. These had hatched on the 23rd February and had just come on to feed. I left them for a week without any attention other than cleaning the screens and had no loss to my knowledge whatever. I killed some for purposes of autopsy and found these had been living and thriving entirely on the midge larvæ and were perfectly healthy. I then set about the task of collecting other food.

Knowing from observations and autopsies that the larva of the *Ephemeroptera* is the chief natural, I made a fine mesh net from a lady's stocking by cutting off the foot and tying a knot therein, attaching the top of the stocking to a stout wire frame 8 in. diameter with a handle about 9 in. long. With this net I combed the short growth of the beds of ranunculus in the shallow waters of the river, working upstream so that all larvæ dislodged would wash into the net by the current. After a number of sweeps the contents were transferred to a bucket of water. I found it was possible to collect many thousands of minute fly larvæ in this manner in a very short time and these I transferred to the stew alive. Care should be taken that the water in the bucket is kept well aerated or all the larvæ will die.

The larvæ established themselves in their new surroundings, and after the first two weeks, from examination by autopsy, I found that together with a few specimens of the midge larvæ, they formed the food of the trout fry. I introduced fresh larvæ to the stew each week. On the 15th May I made a detailed examination of the fry, which were then about eleven weeks old. The majority were $1\frac{7}{8}$ in. long but some were $2\frac{1}{8}$ in. and all were in perfect condition. On the 30th June they averaged $3\frac{1}{8}$ in. long.

Although it is comparatively easy to collect fly larvæ from the river and introduce to the stew by hand, it entails a certain amount of work. I found a much more simple and effective method was the use of fly boards. The flies which use fly boards for depositing their eggs are those which have swimming larvæ and these form the greater part of the trout fry's food. Many thousands of fly eggs can be collected during the spring, summer and autumn by tethering fly boards behind sluices, on bridges, or in any well-aerated water on the river, and the boards, when well laden with eggs, can be transferred to the trout stew. I find these fly eggs come to no harm in transit through the air if kept damp, and hatch in a few weeks.

Those hatched in the late summer and autumn form the food of the trout fry in the following spring; they grow very little during the winter months and are still very minute in the spring. By stocking the stew in this manner during spring, summer and autumn I found the fly larvæ became established and the small area of water was teeming with life the following spring. Stocking with fly eggs should be done during summer and autumn previous to the spring when it is intended to rear trout fry and then continued while the fry are in the stew. I suggest that if streams and stews were made on natural lines, as described above, and stocked throughout the year with fly eggs, by the use of boards, the rearing of trout fry through the early stages would be a simple and inexpensive affair.

The larvæ of the midge can be established by making the stew or stream suitable for the parent fly to deposit her eggs. All that appears to be necessary is to make sure the stew is well aerated, shallow and open to the sunlight. This class of water also favours the algæ which are the food of the food of trout fry. Other life such as caddis, bullheads, loaches, sticklebacks, eels, older trout, and in fact everything except the food of the trout fry, should be discouraged. I have thoroughly tested this method of feeding trout fry for the first three months, over a number of years on a small scale, and have experienced no loss whatever. I can see no reason why similar methods should not be equally successful on a large scale on waters where fly eggs can be obtained.

5   WILD FISH

In April and early May I often have had an interesting hour or two each week in watching wild bred trout fry while they are feeding. Time passes very quickly while doing this—there is great fascination in seeing the efforts of these little fish as they take the tiny creatures that form the food supply. To me they appear to be so eager—so full of vitality as bright eyes search the water, while ceaselessly moving fins are ready to propel the little body in any direction, should desirable food come within vision. I enjoy watching them, but unless you know the habits of trout fry they are not easy to see. Also many little fish of other kinds may be lying along the edges of stream or river. But all species have characteristics that are entirely their own, and when you know of these it is easy to tell them apart. Usually it is the rapid movements of the tail that first attracts attention to trout fry for, even at this early age, the little fish have a pigmentation which blends wonderfully well with its surroundings. Indeed, as you watch, the colouring will change alternately dark and light, as the angle of light is reflected from its sides and back. But what is most noticeable

is the dark barred effect on the sides of the little fish—
Nature's camouflage—just as though the spaced fingers of
a tiny hand had gripped the body and left their imprint.

There it lies, poised in the water of its little holt at the
riverside, all body fins moving and the tail waving from side
to side in a fast, even movement. Now and then the body
may seem to curl to one side as the little fish takes a view to
its rear, and the head may flick quickly from side to side like
a weasel peering from a rabbit hole. A sudden dart out into
the current and quickly back again as the prey is swallowed.
So much for the little trout.

A foot or so away, also poised in the water, is a minnow,
of almost identical size, but note the tail action of this little
fish, and we need look no further for its identity. There is no
fast, even motion to the tail of the minnow, it moves in a
series of jerks, just two or three wags, then a pause, a jerky
pause, and then the action is repeated. Sticklebacks also have
this jerky movement of the tail when lying poised in the
water, but here the pointed head shows a contrast and this,
together with the ventral fin movements, will prove the
identity. These little fish can move backwards or forwards
by movement only of the pectoral and ventral fins, while
the tail is held rigid. Pike can also do this but I do not know
of another fish that can.

Small loaches can also be confused with trout fry—in
fact when they are lying on the bottom they have a very
great resemblance to each other. Often the coloration is
identical and, unless one can see them move, a close study
of the head is the only proof of the species. In movement
it is quite easy to tell the two apart. A loach will shuffle
along on the river bed, but the trout fry will rise quickly in
the water before moving forwards.

These observations of fry resulting from natural regeneration are of value in ensuring that these conditions are chosen
for unfed fry introduced from the hatchery to the river. Of
course the Upper Avon is particularly well-suited to this
operation, but there must be many other trout waters which
would benefit from it. The hatching and early life of fry

living under wild conditions are therefore of interest to all who may at some time have to supervise the introduction of such stock.

Sixty to eighty days may pass while the egg is in the redd and, provided the redd has been kept free of mud and received sufficient oxygen, the egg hatches. The tiny alevin, burdened by its ungainly yolk sac, struggles from the shell and takes cover under a nearby stone. There it will remain for the first three weeks of its life, without moving any appreciable distance; just wriggling here and there, from stone to stone and through the interstices of the heap of gravel, of which the redd consists. But after this period much of its yolk sac has been absorbed, and it finds movement to be easier. It then wriggles gradually towards the outside of the river—to the shelter of the comparatively quiet water under the banks. In some cases the redd is situated near to a bank, in others the redd may be in mid-stream. But whether the trout have spawned in the main river or in a sidestream, whether they are in mid-stream or near one or other of the banks makes no difference, the little trout all make for the edges of the river. Those having farthest to travel, that is, those hatched in a mid-stream redd, will drop down-stream, and may not find the bank until they are anything up to a hundred yards below the place where they were hatched.

The reason behind this migration to the river banks is that the redds are usually located in parts of the water where there is a fast current, and often there is no cover suitable to make a feeding place for fry. It is quite impossible for these little fish to poise themselves in the fast currents which sweep over a trout redd, and it is also impossible for them to see food drifting towards them if they lie in the shelter of a stone, or other breakwater. At this early stage they are neither strong enough nor quick enough to feed in such places, and they know it. So they make for the edges of the river or stream, and there lie poised in comparatively quiet water, yet within a few inches of a little current which brings them their food supply. Such positions are plentiful and are

49

suitable for them during the first month of their fry stage. After this first month, they are strong enough to live and feed in faster water.

Many theories have been advanced regarding migration of mature brown trout. Some say the fish of a river will drift down-stream after they reach a certain age and that this may account for the fact that larger and quicker-growing trout are to be found in the lower reaches. Without question, the lower reaches of a river will produce better conditions and heavier trout than those higher up, but I cannot accept the assumption that trout drift to them from water up-stream. Indeed, from my own observations, I am quite positive there is no migration down-stream by big wild trout and I feel certain that in the natural course of events, brown trout live their lives and die within a short distance of the place where they were hatched or turned in as fry.

Year after year I have run eel traps—stages through which all the water coming down the river has been strained. And during the past twenty-five years these traps, in one year or another, have been working throughout day and night—through every season, and in every conceivable condition of flood and drought. Indeed, I am quite certain that if there had been seasonal migrations of big trout down-stream, then surely I must have caught them, even as I did the eels. Yet, in all these years, with the exception of January and February, when trout return to their former haunts after spawning, I am sure I have not taken more than a hundred trout that have been over $\frac{3}{4}$ lb. in weight. Surely here is sufficient proof that big trout have no tendency to drop down-stream after attaining a certain age. Furthermore, I have questioned other eel trappers whose answers can be taken as reliable. Their experiences coincide with my own. It is the exception rather than the rule to catch a big trout in an eel trap—one is much more likely to get big pike.

So far I have only mentioned big trout—trout of three, or more, years old, and I have stipulated wild-bred fish. In dealing with Nature, one cannot base anything on findings obtained from study which arises from artificial methods. Any stock fish put into a river tend to move considerable distances before they finally settle down. It is only natural that they should, for, on being released from a stew pond, they find themselves free in a strange place, and all start to hunt for suitable environment. Their immediate reaction is to explore, and often enough in a running stream, they choose the path of least resistance. They are unused to running water and varying currents, so the majority drift down-stream.

This can also happen with wild trout bred in the waters of sidestreams which link with the parent water. Some years ago I recollect catching up about 200 yearlings, and two-year-olds, from an adjacent brook. These were put into a deepish reach about half a mile up-stream of one of my eel traps, which was running. During the following three nights I estimated that I took nearly half of these from the eel trap and put them in the reach below. After the third night I had no trouble with the remainder. It has not happened since, as I have been careful to shut off eel traps when doing any stocking up-stream of them. Three nights are sufficient.

Small trout, especially yearlings, even if they are wild bred in the river, have a tendency to move down-stream in the autumn. Here, I think, is a natural desire to migrate to more suitable feeding grounds. In this case, it is a movement from deep water to that which is more shallow and therefore more productive of the food required. These migrations occur in the autumn and often coincide with the seaward run of the eels. Now eels usually run with the first autumn flood water—with the dirty river conditions we are accustomed to having after a period of heavy rains. And this dirty water condition is, I think, the answer to why the small fish choose to drop down-stream.

In the natural course of events, these little trout would have hatched from eggs laid on the shallows, or perhaps they were introduced as fry to the upper parts of the reach on which the eel trap is situated. Their early life would be spent on the shallows but as they increased in size and in strength some of them would drift down-stream to the deeper water just below the shallows. Here conditions have been favourable during summer and early autumn—water has been clear and food plentiful. Then come the floods and the dirty water causes adverse conditions.

Usually, eel traps are constructed where there are controls across the river, a set of hatches, a weir or the like— in fact, they are at the down-stream end of a long reach of water where the river course and banks are man made. Indeed, it is a place to which the natural course has been diverted, where the water has been led into a trap to build up heads for milling and irrigation. These reaches are just places where mud and other filth collects from the natural course up-stream and with the rising water and additional flow, this filth is moved to mingle with the water and be swept towards the sea.

The little trout find that they cannot feed and that respiration is difficult. Some may go up-stream to their former haunts in the shallows, but many drift down with the faster current in the hope of finding a place where they can live. In this they are successful. After passing through the controls, the water regains the natural course and in most cases this is both shallow and gravelly. Such water, aerated as it is in its fall from the higher reach, is quite suitable for the young trout and there they find a home and a sanctuary. These shallow, natural conditions serve them during the winter months and by the following autumn, as two-year-olds, they are capable of living in any class of water.

When catching up trout for stripping purposes it is very useful to have some idea how to tell the difference in their sex when in the water. There are times when one sex or the other is needed, and when it is not possible to catch them both at one and the same time. I speak mostly of occasions when trout are spawning in sidestreams, or in other places where you can watch their movements after they have had warning of your approach or of your intention to catch them.

Often, though both fish seek shelter under a bank, a tree-root, in a weed-bed or some other place of concealment, they do so individually, and choose parts of the stream that are widely separated. You may want one hen fish to complete a quota of eggs, or perhaps a male for fertilisation, and it is both helpful and time-saving if you know where each has gone. Otherwise the likelihood is that you will catch up the wrong one and allow the other a chance to escape.

At spawning time both male and female trout lose a lot of the shyness that is so habitual at other times throughout the year, but the sense of self-preservation is dimmed much more in the female. I have found, by studying the reaction of the paired fish to my approach, that I can, by this one observation, invariably distinguish the sexes. You may perhaps come across a pair which are travelling up-stream during the hours of daylight—a pair which have not started to spawn and which are carrying their full quota of eggs and milt. The female leads with the male just behind her. You will find that it is *he* who first becomes aware of your presence, and that it is *he* who makes the first bolt for cover.

When fish are on the redds, the male sees, or senses, your approach much more quickly than the female and he will immediately abandon her to whatever fate there is in store. It is something I cannot understand. The fish may have been paired and living together for several weeks previous to spawning, yet the male has no sense of chivalry—no regard for the well-being of his mate. Occasionally, she also is alarmed at once, but it is difficult to say if this is due to knowledge of your presence, or through the sudden departure of the male. But, in most cases, the female will stay on the redd for some moments after her mate has retired. Sometimes she will realise the danger after a short while and will then go to shelter, but more often will remain in the hollowed gravel until her mate has plucked up sufficient courage to return. However, should both fish move away from the redd, it is always the female that is the first to return to it.

But, in addition to these habits caused through the difference in temperament, courage, instinct, or whatever can be the cause, if the water is clear and a good view of the fish can be obtained it is usually possible to determine the sexes by the contrast in the shape of their bodies and heads—the latter being the most reliable. A female has a certain fullness of body when she starts to spawn, but as her eggs are laid so this fullness departs to leave the two fish with a somewhat similar outline. But the head of a female trout is always much more shapely than that of the male, and this is very noticeable when the two are together at spawning time—the nose is rounder, the head smaller, and the edges of gill covers less sharply defined, being set nearer to the eyes. There are times when I have to take fish with my hands—from under banks, or other places, or perhaps from a container where I have both sexes ready for stripping. It is then no use relying on habits or on sight—one must depend on sense of touch. I have found that there is always a certain firmness of any part you may touch of a male, which contrasts with the soft and velvety feeling you experience when touching the female.

One curious incident gave me further reason to think that trout lose their customary sense of self-preservation during the period they are spawning, and indeed at other times, if in some way two males have a difference of opinion. For many years, with the exception of three seasons during the last war, I have caught up sufficient wild trout at spawning time to provide about 40,000 eggs for artificial incubation in the hatchery. One easy method I find is to use large-sized pike traps to catch the spawners as they run up some of the small sidestreams. One year I was using this method in a small brook which always attracts spawning trout. I had the trap set some distance up from the river and as I walked up the stream bank towards it I saw what I thought to be a pair of exceptionally fine trout. The water was discoloured through heavy rain the previous day, but I could clearly see the two fish drifting, tail first, down-stream. Every now and then one of them showed a dorsal, or tail, fin above the surface, and each milled around, about, and over the other, with occasional dashes up-stream for a few feet. I knew they were not spawning as the stream-bed at that point was muddy and unsuitable.

I was anxious to secure these two fish, the hen would probably produce some 1500 eggs, and so I tried to drive them up-stream so that they would both enter the trap, twenty yards above. But though I made myself very conspicuous neither of them moved apart, nor would they bolt up-stream more than a few feet. I tried several times, but each time they drifted back again and all the time were playing and milling around each other. I had with me a very short-handled landing-net that I use as a keep-net into which to transfer trout from the pike trap, to be ready for stripping, and I had on a pair of rubber thigh boots. Chasing the trout once more up-stream, I stepped into the water behind them.

Almost at once they drifted towards me, and I held the landing-net ready submerged in the path of the one I thought might be the hen fish. Backwards into the net it came, and I quickly lifted it clear of the water. A quick glance showed that I had been mistaken in sex. It was a cock fish of well over 2 lb. The other fish, which I now took to be the hen, rushed up-stream for a few yards and then, to my astonishment, drifted tail first towards me. I had nothing into which I could transfer the one I had in the net, so I dipped net and fish quickly down behind the other one. I had expected the netted trout to bolt out immediately the net ring was under water, but to my surprise, instead of the captive fish bolting out, the free one turned quickly round, as though attracted by the disturbance, and with jaws agape, plunged headlong into the net. I had them both, but both were males almost identical in size.

Shortly afterwards I examined the pike trap. In it were two hen trout which could have been the partners of the two big cock fish. Possibly they had run up the stream together, each big male with his respective mate. In some way the entrance to the trap had been partially blocked and, though the females in their anxiety to get to the redds up-stream had managed to force their way inside it, the males had failed to make an entry. And so it would seem that in their endeavours to follow their mates they had become jealous of each other—one probably suspecting the other of spiriting away the lady of his choice—and so jealous had they become that in the end they had forgotten their mates and, when I appeared, were still fighting it out, up and down the brook, half in and half out of the water, regardless of love or life.

The subject of a trout's reaction to approach and capture raises the old question of "tickling". For hundreds of years trout have been caught by bare human hands. Possibly the idea originated through men studying the feeding habits of bears, which often fish in a manner that is similar. The method has been described as tickling. It has been said by many of the older writers that the gentle tickling movement of the human fingers beneath the body of a fish will lull it into a kind of coma, so that it becomes unaware of what is happening; that in this condition it will rise to the surface of the water on the tickling fingers, when removal to the bank is mere child's play.

The fact that fish are cold-blooded perhaps accounts for their docility, for my experience in human nature is that it is often the tickler, instead of the victim, who reaches a state of coma. Yet it is quite true that trout and other fish can be taken from the water with bare hands, and I expect that during my life I have caught many thousands by this method. It is also true that they have no objection to being touched by moving fingers. But I do not think they are unaware of what is happening or that they are in any way tickled silly, or pink, as someone once thought. It is just that they have a wrong perception of what is taking place.

The movement of fingers beneath the body of a fish will cause it to rise in the water; continual gentle movement will bring it to the surface, where it can quickly be swept out on to the bank. But it is not because the fish is in any state of coma, as one soon discovers if any quick movement is made in getting the fish to the surface, it is just because the trout has no weight in water, and the movement of the fingers, however gentle, is sufficient to propel it to the top of the water. Neither is it because the human fingers have any peculiar quality of their own, for a trout can be raised equally well with a short stick.

Although most people who write about trout tickling explain that the procedure is carried out in shallow water—beneath the river or stream banks, under tree stumps and roots and in the river vegetation—they seldom explain the fact that, while in such places, the fish have cover over them, and therefore feel secure. Actually they are in one of their many holts where they spend many of the daylight hours. These holts are often used by several fish of varying size and in many cases the trout are lying side by side, across and above each other, so that one is continually touching another. Small fish may wriggle beneath the body of a larger one and so pass to the other side. The movement causes no alarm, they are accustomed to it, and so when human fingers, chilled to the same temperature as the water, slide beneath their bodies, they take no notice.

I often have to take trout with my hands during the spawning season—not only to take them from the water but to handle them with care. Many of the fish I take are mature females, and their quota of eggs need little pressure to extrude them. It is then necessary to operate with both hands. Though tickling trout may sound romantic, actually there is no necessity to tickle at all. Gentle movement of the finger tips acts as a lever, and will lift the fish high enough from the river bed to allow the hands to slide to positions at head and tail, when a firm grasp can be obtained. Trout (and for that matter any other fish) have great difficulty in struggling forcibly if the mouth and gills are held closed, and if a grip is also held on the tail part of the body, the fish is powerless. In such manner a trout can be held firmly, lifted from the water and carried, without in any way damaging it.

And so the trout can be taken from his holt, unaware that the gentle movement beneath his body is other than caused by natural events. He lies quiet in the shaded light while what he thinks is part of one of his brethren moves gently to his head and his tail. He may wriggle a bit, unwilling to be jostled from his position, but until the firm grip of the human hand fastens and traps him, he has no idea anything is wrong.

6   FLY AND NYMPH

To me a river is like a picture painted by a great artist. There is far more in it than meets the eye of the casual observer. One has to look at it from many angles—notice the light and shade, look beneath the surface, and see how the brush has merged the colours of so many things into a common background. Only by being able to appreciate what a painter had in his mind can one enjoy to the full what he has put on his canvas. The pleasure comes, not so much with the background presented by the water in the river course, or from the scenic effect of its surroundings, but by being in a position to look deeper and enjoy to the utmost the life of the many aquatic creatures which live in those liquid depths. It is only when, under the microscope, you start to examine some of the animals of a river that you can fully value the beauty of shape and colour that was given to them in their creation, for the magnification reveals that even the most insignificant of river life has a coloration and symmetry that far surpasses anything which lives in the open air. Throughout the seasons from January to December the life in the river is ever changing; the picture captured to-day has gone to-morrow—there is something new to take its place.

There are people who think fish have no powers to discern the different colouring of their food—of insects especially. To that question there is no possibility of an answer, but what I would like to know is, if fish have no means of distinguishing between one pigmentation and another, why it was thought necessary to provide such colours in the first place. We are apt to judge wild creatures by our own senses, by what we see with our own eyes, and by what we are capable of thinking. In this I feel sure we are wrong. We like to think we are the most advanced of all animals, but are we?

After all, we can see the colouring of these tiny insects when our eyesight is assisted by magnification, and I for one think it possible that the eyes of fish might be like ours are, when aided with the microscope. Why fish are deceived by the crude imitations we offer them is a fact I can never understand, for none, not even the most carefully arranged artificial, can have the same detail and exact colouring of nature. I think it must be that some movement is imparted by the action of the water on the dressing and that this suggestion of life is sufficient to blunt the senses of fish to their more common ones of being able to discriminate by habit and pigmentation.

That a trout will at times feed exclusively on one particular species of fly, though there is ample opportunity for him to take others, is a fact that is well known. To the naked human eye there is little to distinguish one of these from another, but when the assistance of magnification is enlisted the difference is at once apparent. Immediately we can see features that are contrasting. The more one studies these river animals, the easier it is to tell them apart. Each species has its own individual colouring and characteristics, both in the water and out of it and I feel sure that if these creatures were *our* principal source of food we would very soon be able to tell each of them at a glance—we would soon know which was of the most value—which was most suited to our palate at a certain time.

A dry-fly river can be made or marred, not by the quantity or quality of the trout it carries but by the numbers of suitable flies it will produce. It is essential that good hatches of fly should be forthcoming on any river where dry-fly and up-stream nymph fishing are the only methods employed. Though sparse hatches will bring the small fish to the surface it needs a really good hatch of fly before the larger trout consider such food worth waiting for. It is well known that the winged flies taken by a trout form but a small percentage of the fish's food. Trout can live very well throughout their lives without taking a single fly. Even on rivers such as the Upper Avon many of the large fish never rise at all unless the hatch of fly is exceptional. Then, confident of a fair return for their labours, they rise steadily but cautiously and eventually cease when the hatch of fly falls below their minimum, when it is no longer worth the effort to rise. Sometimes, however heavy the hatch, they lie dormant, leaving it to the smaller ones to have a hearty meal.

Sport therefore depends on fly, not of one species but a dozen or more different families, flies for every month of the season; and flies in such quantities that all the best feeding positions in the river will be occupied by the heaviest fish in the locality. We know full well that we cannot make the flies hatch to order, but if the insects are in the river in quantities, then there are times when weather and water conditions will be favourable for them to hatch in great numbers.

Before the war I tried various successful methods for increasing fly life. Gradually I built up a good stock. Trout were plentiful and good bags were obtained. Then came the war and much of the work I had done was ruined through pressure of other duties. Still, 1946 proved a very fair fly year and I knew that, with careful nursing, a year or two would see again the grand hatches of pre-war days. It was therefore with a sad heart that I saw the postponement of my plans brought about by one of those disasters with which a river keeper must contend. This was flannel weed and flood.

Flannel weed, as it is most commonly called, breeds only in the latter months of the season, usually starting about the end of June when the temperature of the water has risen sufficiently to encourage its growth. Hot weather, slow running water, lack of aeration, absence of other better-class weeds, all tend to allow it to spread, and often by the end of August much of the river bed is covered. Though very bad, the growth of flannel weed in 1946 was no worse than in other years I have known, but the great trouble came with an exceptionally wet period in November. Great carpets of the emerald green growth had spread over the beds of many reaches of the Upper Avon. Fly life of all species and in all stages of life were living amongst the hair-like fronds and one could, if he so wished, pull out a great tangled mat of the weed with a rake and find many thousands of creatures struggling to get free. The weed might perhaps roll as it was being dragged to the bank, so trapping every insect inside as successfully as though in a blanket. I pulled out many such lots and several times put the rolled mass back into the river just to see if the creatures trapped within could make their escape. I examined some of these many days afterwards: the insects were still there and many were dead.

In the natural state the weed remains as a carpet on the bed of the river and starts to decay about the middle of September. The filaments gradually rot away and the carpets break into fragments. By the end of November much of it has turned into mud. But 1946 was an exception. Rains came in early September and the river rose bank high. The flannel weed remained in a fairly healthy state. Then in November came more heavy rains. The river, already comfortably full, received more surface water, and in consequence all the available controls had to be opened to prevent flooding. The river course became a torrent.

The flannel weed, though it had commenced to decay, was still in the form of a loosely woven blanket and as such it moved off the river bed. Areas of perhaps thirty square yards, aided by the drag of the current, lifted off the bottom and rolled into great masses. Though they were too heavy to float, they were swept down-stream. All would have been well had it been possible to keep the water confined to the river banks and to pass it through the controls, for much of the weed would then have been broken up in its passage through the sluice gates. Unfortunately these were in-adequate. Water and weed swept over the river banks into the meadows at many points, and the heavy vegetation was cast into great heaps many yards from the river course. To make matters even worse the water receded even faster than it had risen and within forty-eight hours was once more in control between the river banks. But the damage had been done.

Stranded amongst the grass and other meadow vegetation were millions of water-bred insects, and the masses of flannel weed were just great death traps. I pulled apart many of the heaps and was appalled at the mortality that had occurred. Fly life of all species and in all forms of their under-water existence were caught in the maze; and, in addition, snails, shrimps, crayfish, bullheads, loaches, minnows, water boatmen, beetles, worms, frogs, lampreys and leeches, all were hopelessly stranded and, apart from the frogs and perhaps the worms, all were doomed to die. I know many of the chalk stream flies in their nymphal and larval stages and could see at once that the Blue-winged Olives were particularly numerous in their early larval stage, and this I could readily understand, knowing full well the ponderous underwater movements and the poor swimming qualities of these creatures. I could imagine them trying to swim from the moving masses and getting hopelessly entangled.

Though there were large numbers of Mayfly larvæ present I had little fear that the species had been badly affected, as I knew that the majority had already taken up their winter quarters deep in the river bed and banks. Stone fly larvæ (little creepers) had suffered badly. These poor helpless creatures could not even attempt to swim from the trap and, together with numbers of caddis and free-ranging sedge larvæ, were swept to their death. The swimming group of fly life such as the Olives, Pale Wateries and Iron Blues, though numerous, were not in the quantities I had expected, and I think it possible that many had managed to swim clear as they were being swept along, but I knew as I examined each mass of weed that the stock of small fly was ruined and that the season of 1947 would show a great shortage. And so it proved.

Mayfly were numerous, but no other flies appeared in any quantity until late in the season. Later, October and November brought again the large hatches of small flies I had been so accustomed to see during the summer months, the flies having undoubtedly hatched from the eggs that were laid in August, September and October, 1946. Many of these eggs at the time of the flood were still adhering to the vegetation or stones on which they were deposited and were not moved by the flood water; or were in such tiny form that they managed to escape the grip of the flannel weed. The absence of Blue-winged Olives was particularly noticeable as these flies are usually very numerous on the Upper Avon.

Flannel weed is always a curse wherever it is found, but in times of early flood it is a means of ruining the following dry-fly season. As far as I know, no one has yet devised a method of controlling it once it is established. Well-aerated water will prevent its growth, and dense masses of the good-class vegetation such as rununculus, watercress celery, and so on, will suppress it; but its removal from the river has been found to do harm by the unavoidable destruction of insects inhabiting it.

Experiences of former years should remain vividly in the memories of fishermen and be a guide to their choice of water when thinking of fishing early in the season, but each year I am asked the same question by fly fishermen, often by the same ones year after year: "Where do you think is the best place?" April is seldom a good month on south country streams, but as dry-fly fishing must depend on insect life on, or in, the surface film of the water, so in thinking of rising trout one should bear this fact well in mind. The great thing is in knowing where flies are likely to be hatching so early in the season and where they may be appearing in sufficient quantities to interest the fish, for it is there you will find your sport. Occasionally there can be really first-class hatches of flies of various species. Iron Blues may appear in great quantities. Black Gnats may hatch from about the meadows and some fall on to the water. If the season is a late one there may still be hatches of Grannom. But the main flies of the spring are the Olives. Of these there are usually two species, the large dark spring olive and its smaller cousin, a fly that is almost identical in coloration.

With these two flies a third may appear, a fly much lighter in appearance both in body and in wings, which fishermen take to be the light olive that usually is well in evidence in May. Light olives may possibly be hatching in small numbers early in the season, but I believe the majority of the light-coloured flies, which hatch in April to be a species of Spurwing. I hope in the next few years to be able to confirm my discoveries of the past seasons, and put the matter beyond doubt.

Though a few spinners may be floating in or on the surface film, it is upon Olives that the fisherman should base his thoughts, for these duns will be the principal flies upon the water. The larvæ and nymphs of these flies have a preference for shallow and well-aerated conditions, places where the green fronds of ranunculus are thrusting through to spread in great tresses on the surface. Here the fast-growing weed bunches are transforming the gravelly reaches into a series of runs and glides, of little pockets and eddies, a sanctuary and feeding ground for immature Olives. It is amongst the well-aerated foliage that the nymphs have been feeding during the past few weeks, for there they have gathered from their winter quarters in the gravel, and it is from the weed-beds that many of them will hatch. In the quickly-spreading vegetation of the shallows are collected what will be the flies of early spring, and none knows this better than the fish.

Weed-beds also have become the main feeding ground for trout. Whereas after spawning, and particularly in March and early April, the fish feed in more slowly-moving water upon caddis, snail and shrimps, as April advances fly life becomes their main source of food. They congregate where it is most likely to be obtained. On a well-stocked water, every little run has its tenant, some of them have more than one, all eagerly hunting, watching and waiting, to take toll of the insects as they move amongst the weed strands.

During this early part of the season flies hatch slowly. The water is too cold for them to leave it quickly. Often the duns will float for a long period on the surface before taking flight and with the swiftly-moving currents may be carried many yards down the stream, to present opportunities that trout find hard to resist. So in the early season it is to these shallow reaches of the fishery that the fly fisherman should make his way. He would be advised to forget his experiences of summer evenings and autumn days, and waste no time in watching the deeper reaches where, then, fishing was productive. Let him go where the river is awakening to the call of spring, let him watch carefully the ripples and runs, the eddies and pools amongst the weed-beds, for that is where there are both flies and trout, and as the shallows are bathed in the warm sunlight of early afternoon, he should find plenty to enjoy.

The winter of 1948–49 was an exception, a queer winter following a queer summer. Nowhere was this more pronounced than on and near the river, for scarcely a day passed between the end of September, 1948, and 1st March 1949, without a hatch of small flies being present on the water of the Upper Avon. Though it is nothing unusual to see hatches of such flies as the large and medium Dark Olives, and occasional appearances of Iron Blues and Pale Wateries in winter, anyone writing or speaking of the presence of the Blue-winged Olive (*Ephemerella ignita*) at Christmas was held to ridicule, and suggestions made that the observer had probably mistaken the fly, it being in fact a large Dark Olive.

Yet in the winter of 1948–49, not only at Christmas, but all through December, January and February, I recorded many good hatches of Blue-winged Olives, not just one here and there but hatches in sufficient quantities to gladden the heart of any fly fisherman had they appeared during the fishing season. I gave these flies more than a casual glance to determine their species, for, with the aid of a microscope and notes I have made in the past, I proved them definitely to be *Ephemerella ignita*, the fly known throughout the fishing world as the Blue-winged Olive.

One day in particular—4th January 1949—I counted forty-two flies on about an acre of water; I know most of the river flies at a glance and the B.W.O. has characteristics which to me are unmistakable. Undoubtedly these were all B.W.Os. It was midday. Trout were spawning, but grayling had every opportunity to rise if they wished, yet not one of the flies was taken by a fish. The insects just floated unconcernedly along, in some cases over a hundred yards, before taking flight. This, incidentally, is the procedure with most of the flies which hatch during the winter months. It would appear that the colder air necessitates a longer time for the wings to dry sufficiently for flight. Now strangely enough, though I saw several really good hatches of B.W.O. during the 1948–49 winter, I did not see a single female spinner come back to lay her eggs, neither did I see one about the countryside. What is stranger still, of the forty or fifty flies I caught and examined in the dun stage, not one was a female. About the meadows, I saw scattered groups of a few dozen male spinners dancing during two warm afternoons in January, so it would appear that weather conditions allowed these flies to cast their dun covering, and so change to the perfect insect.

Conditions during the summer of 1948 were very unfavourable for all ephemeropteran flies. It was a low-river year—a season of water of high temperature which bred flannel weed and produced a film on the water which presented a barrier many nymphs found impossible to penetrate. Flannel weed appeared early in the season in the Upper Avon, and this not only sapped the oxygen content of the water during the night, but at all times fed on and stifled much of the minute life which makes the food of flies. And it would seem that unless the larvæ get sufficient food they will not mature—their nymphal stage comes much later—and hatching is delayed. This, I am sure, was the case in 1948—the growth of crawling larvæ and nymphs, such as the B.W.O., was affected even more than that of the swimming group; they were starved, and in consequence did not hatch at their proper time. The absence of B.W.O.s in the Upper Avon was particularly noticeable during June and July, 1948, months when these flies usually hatch in great quantities. Although I have since found about thirty per cent. of Blue-winged Olives hatching in January to be females I have not seen a female spinner and I doubt if eggs are laid by this species during the winter.

In the second half of April the Hawthorn fly appears and often this ungainly black insect is evident in some quantity down the Avon Valley. The Hawthorn is a land-bred fly, and it would appear that there are two species, which vary considerably in size. In both instances the female is much larger than the male, but otherwise the characteristics and colouring of both sexes and species are the same. Eggs are laid in various places about the country-side, favoured spots being heaps of rotting vegetation, roots of grass tufts under the bark of decaying trees, and in damp ground. In such places the larval and pupal stages are passed, the whole cycle taking one year.

The actual existence of the Hawthorn as a flying insect is brief, and is principally for the purpose of mating and egg-laying. Large gatherings of the flies can sometimes be seen dancing and gyrating around hawthorn bushes, and this may have led to the belief at one time that these flies hatched from such bushes. Though they favour a river valley, they may also be seen in places many miles from water. The Hawthorn could well be described as a large black gnat; the female of the larger species is about $\frac{3}{4}$ in. long, with a plump, shiny black body, and two large diaphanous wings, which lie flat on the back of the insect when at rest. When in flight, the long black legs of the insect hang downwards—a characteristic of this particular fly which is unmistakable.

Owing possibly to the mild winter of 1948–49, which

allowed the larvæ to thrive and pupate, exceptional numbers
of the Hawthorn were to be seen in the Avon Valley during
the last week of April and the first week of May, 1949, and
in this time many fell on to the river and were a great
attraction to the trout. During this fortnight we had some
queer weather, and one afternoon I remember a freakish
storm which came from the south-west. The morning had
been warm and windless, and about the meadows I had seen
many hundreds of Hawthorns dancing, and mating. It was
about 2 p.m., when suddenly a great wind swept across the
valley. The calm of the river was quickly transformed into
a miniature sea, with waves, six inches or more high,
sweeping up-stream and slushing under the banks. There,
riding the waves like black dinghies in an ocean, were scores
of struggling Hawthorn flies, males and females alike,
which had been swept with the wind on to the water.

I watched for a moment and then I saw some bubbles
behind a wave, then others, and within two minutes, heads,
dorsal fins and tails were showing repeatedly in trough and
wavecrest while the struggling black flies disappeared
beneath the water. Never before or since have I seen a rise
of trout to equal the few minutes that followed. I should
think every trout in the water took part in it, and in that
short period of less than ten minutes, the fish must have
taken nearly 1000 flies from the water within my view.
That, however, was an exception. I might never see such
an occurrence again. But there can be no question that trout
are fond of the Hawthorn fly and regardless of weather
conditions will take the insects when accidents bring them
within reach. They are very awkward in flight, and appear
to fly aimlessly about the valley. While mating, the paired
flies fall to the ground, or perchance on to water.

I was amused when watching some last year during a
warm afternoon in early May. Mating was taking place all
around me, and every now and then a pair would fall together
on to the river. I often remark that Nature knows no chivalry
and this was borne out forcibly. As the paired flies struck the
water, the male scrambled on to the back of his mate, and

used her as a take-off position to regain flight. As he rose into the air, so his weight pressed the female deeper into the surface film, where she was hopelessly trapped, and where she remained until her life was ended in the jaws of a fish. Almost immediately the male found another mate, and the last I saw of them was their falling into the grass of the meadow. No, Nature knows neither chivalry, nor remorse.

7 MAYFLY AND AFTER

THE early appearance of a few Mayflies is often a good indication that a season of big hatches is approaching. Each year I see one or two about long before the proper season starts, and these odd flies are usually males. As the season approaches, a dozen or more may appear daily, and then numbers gradually increase as the time for the main hatch draws near. But though a few dozen of these flies may appear daily and float on the water, a week or more may pass before any are taken by fish. Then suddenly, as though moved by some general impulse, the insects start to hatch in great quantities, and trout begin to feed.

It is hard to say just what causes this mass initial hatch, but I feel sure that weather and water temperature both play a big part. The Upper Avon is often quickly swelled by surface water draining to the valley after rain, and this rising of the river may come twenty hours, or more, after the rain has fallen. In 1948 and 1949, the big hatch started with a rising river—a river discoloured and warmed by heavy rain the previous day. The hatch started about two o'clock, and 1949 was just a repetition of 1948. In both seasons, this initial hatch of flies continued for about three hours, and in this time many thousands of Mayflies appeared.

I get a lot of enjoyment in watching trout when they really come to Mayfly at the start of a season. It amuses me to see the little hesitancy, a cautious rise or two, and then the

madness as they let themselves go, rising with a total disregard of everything and everybody, and abandon themselves to an orgy of feeding. But with these initial hatches
of Mayflies everything is so sudden. One moment an odd
fly floating here and there and the next the surface is one
great scene of activity, with Mayflies hatching as if from
every square yard of the river bed. There they are in their
hundreds, some floating with the current, quivering as they
separate their wings, some fluttering along the surface in
clumsy attempts to leave the water, and others airborne,
flying to the shelter of the surrounding countryside. It would
seem also that the sudden appearance of these insects is just
as much a surprise to the fish as it is to me. And I think that
for a short while the trout cannot believe the evidence of
their eyes nor the good fortune Nature has presented for,
though the flies are hatching, fluttering and flying within
easy reach, they remain for a few minutes unmolested. Then,
almost as suddenly as the appearance of the flies, comes the
movement of the fish.

A Mayfly disappears in mid-stream as the head of a trout
shows above the surface, another sinks into a dimple as a
fish rises under the opposite bank. Widening circles show
where yet another fish has risen up-stream, and then a
splash just below as a little trout hurls itself clear of the
water in pursuit of a fly fluttering above the surface. Then it
seems that this splash has acted as the dinner-call for all at
once, the river is aboil with rising fish. Splashes now take
place everywhere; waves surge here and there; and the sunlight glints on the sides and bellies of little trout, as in their
excitement they throw themselves clear of the water to catch
the moving flies.

Trout of all sizes are lured to the surface, and forget all
thought of self-preservation. They lose all sense of direction,
and rush madly about the river. To these fish it is feast time,
and they are like a flock of lambs in a field of clover. There
they are, irresponsible, happy and carefree with food in
plenty just for the taking, and for a while they enjoy
themselves in the madness of Mayfly time.

Time and again the Mayfly season has proved wrong the old saying that it is the duffer's fortnight. Granted there are some days when the catching of trout is comparatively easy; there are also periods when even the experts are forced to admit defeat, and watch hour after hour while the fish take the natural flies with eagerness and a total disregard of human presence or artifice. But it is sometimes to be noticed, also, that on these days of discrimination the fish avoid the natural fly if it is a dead one, and I am indeed doubtful if one could kill trout at such times if one mounted a freshly hatched dead fly on a hook, and fished with that. I am sure that, in general effect, many artificials are good representations of a dead Mayfly, and had trout been accustomed on such days to taking flies that had died, then the imitations would have been successful. And here is the point of the matter.

In one season the Mayfly season on the Upper Avon started in quantity with a rising river. For several days previously sufficient flies had hatched to whet the appetite of the trout and they had taken them freely, so that when the big hatch came on, the fish were already accustomed to seeing these big flies on the water, and to taking them. The first big hatch, therefore, met with great response and most of the trout in the river came on to rise madly. But with the cloudy water it is reasonable to assume that the trout's view of the hatching flies was somewhat obscured. For two or three days they were quite easy to deceive with an artificial. Then as the water regained its crystal clarity, and sank to its former level, the fish became " choosy." Though Mayflies hatched in many thousands the fish took only those that showed some movement and completely avoided all dead naturals and artificials. Though so many flies had hatched, the weather was apparently unsuitable for the females to return for egg-laying, and so far during the Mayfly season the trout had had no opportunity for seeing dead or dying flies in any quantity.

Then came an afternoon when the spent gnat were on the water in thousands. For a little while the trout took only those showing some signs of life, but gradually the dead flies were also taken. It seemed as though the trout had only just realised that the dead flies were quite as good to eat as the live ones. The artificial was again accepted readily, dead naturals were taken without hesitation, whether spent or freshly hatched.

There was great talk amongst fishermen about the educated trout in the fishery and about the uncanny manner in which the trout took the natural flies and refused all imitations. Various reasons and theories were put forward to account for it, but I don't think any were near the truth. For it seems to me that the answer to the problem is the simple fact that until they have had the chance to see May-flies dead upon the water in quantities, the trout know these big insects *only by their combined size and characteristic movement.* Not one freshly hatched fly in a hundred remains still on the water. The occasional dead fly passes the trout without attraction, not because he is educated and thinks it might be an artificial, but simply because he is looking for a big insect with movement; and for this same reason he lets an artificial pass over his head. One has to remember that the Mayfly season is but a short period each year, and that the larval and nymphal existence of Mayflies is spent in places where they cannot be seen by trout. It is therefore very doubtful if trout can remember from year to year that Mayflies are good to eat. Each season he has to have the same lesson and learn the fact anew.

For some time after the appearance of the first Mayflies of the season, the fish will allow the insects to pass over them regardless of whether they are alive or dead. They have forgotten the previous year, and for a while cannot recognise the insect as something edible. Then, after a time, the moving insect becomes irresistible, every one is taken and is found to be good. Then to the river come the female spinners. Size is there, and an even greater movement as the insects dance and dip to the surface. The fish, in their frenzy of excitement, leap into the air to catch them, but soon the bigger fish realise that if they wait patiently the insects will fall to the water and drift towards them; so they take up positions and wait. They see the insects flutter and then remain still. Another lesson has been learned. They find that the dead flies are good to eat—that movement is not essentially an indication that it is indeed an insect.

So it seems that if trout are to be taken successfully with artificial Mayflies, then the representation must be made in such a manner that it appears to the trout as a moving insect. A very lightly dressed fly with long hackles tied in to give a buzz effect—hackles sufficiently long to make a circular ripple around the body of the fly, and to ensure that it sits high on the water. Bright blue cock hackles would merge with the sky and be invisible after giving the effect desired. Everything is worth trying. I hate to be beaten by trout.

One evening I had a very interesting half hour watching Mayflies. I was standing close to a long, well-trimmed privet hedge, some four feet high. Over this about a hundred male Mayflies were dancing. At intervals a female would appear from the surrounding countryside, fly into the assembly of males, mating would take place and then I watched as she made her way, with her characteristic undulating flight, to the river a hundred yards away to commence her egg-laying. For some time this continued, and then I could see a storm approaching. As the sun disappeared behind the storm cloud, so all the male Mayflies folded their wings and dropped quickly downwards, taking shelter in the privet hedge. I examined some of them as they clung to the undersides of the privet leaves, and, to all appearances, they had little life in them. None would move, though I touched several.

Nowhere could I see either male or female in the air, but after a few moments I noticed a female about fifty yards away, coming in the direction of the hedge, one which was very anxious to lay her eggs and intended to do so regardless of weather. Her immediate necessity, as her instinct prompted, was to find a mate, and as the sleepy-looking males clung to their refuges, so the female hovered and coursed up and down the hedge, above them. Backwards and forwards, the full length of the hedge she went four times, without getting any response. I shook the hedge to see if any males would take flight, but all remained where they were. Then down she dived into the privet. There she paused for about half a minute out of my sight, and then rose into the air followed by half a dozen males. Mating quickly took place and soon she was on her way to the river to lay her eggs.

What astonished me most about this incident was the fact that this female knew there were males hidden in the privet hedge. She was a long way away when I saw her approaching and as she was then high in the air, and quite clear of any tall vegetation, it is possible she had travelled still farther, so there could hardly be any question of her being able to see the males when they dropped for cover. She was the last female I saw until the storm had passed, and sunshine once more brought the males out to dance.

This seemed to me to be a case of a female who was becoming unable to retain her eggs. It is possible that she had come for a considerable distance, and had encountered no assemblies of males en route. Her act in searching up and down the hedge was probably one of desperation, but whether the males saw her and completely ignored her I do not know. In her flight above the hedge it is possible that she caught sight of one, or more, of the sheltering males, and her appearance amongst them quickly woke them from their stupor.

It is generally assumed that if weather conditions are favourable for reproduction a female Mayfly will change from nymph to sub-imago, cast a further shuck to become the imago, then mate and return to the water to lay her eggs and die all in the course of about twenty-four hours. But do they? I had reason to doubt this some years ago when I saw several Mayfly spinners leave the water and hasten away into the countryside. I was curious to know what they were going to do so I followed an individual. I had seen this fly dipping and egg-laying in the usual characteristic manner—in fact, I watched her long enough to be sure she was indeed depositing eggs, for I could see them on her oviducts each time before she dipped to the water. In all she must have dipped quite thirty times.

After travelling across the meadows for about two hundred yards she fluttered into the long grass, and there I found her. I was curious to know if she had indeed deposited all her eggs, and had just dropped—spent—into the grass to die. Taking her home, I carried out an autopsy. What I found surprised me, for that insect had only laid about one half of her eggs, the others were still within her body. This was so interesting that I decided to follow it up, so as to find out if this was an exception to the general rule of egg-laying. Since then I have seen the same occurrence very many times, and have examined large numbers of egg-carrying females under the microscope.

In the past I had been puzzled by the fact that some females would fly straight through an assembly of males without receiving attention, though they were of the same species, and would go to the river and commence egg-laying. On examining some of these I found they were not carrying a full load of eggs and in fact, that they had already laid a certain number at some time previously. It would seem that the remainder of the eggs were already fertilised, and that this was known by the males. I caught many other females after watching them mate, and in every case these carried their full quota of eggs. I gave consideration to the possibility that this break in egg-laying might be due to river conditions or to atmospherics, but on evenings when to me everything seemed ideal the same thing occurred. I watched females mate and go to the river, watched them lay part of their eggs, and followed some as they returned to the shelter of the countryside. Whether they leave the water a second time after egg-laying I have not yet ascertained, but I think it quite possible that they do.

As a result of these observations, much of what had puzzled me in the past then became clear. I have seen many Mayfly seasons start with a big hatch of fly, and though I have seen occasional flies egg-laying I have never yet seen a big fall of spent gnat (spent egg-layers) the following day. If it were true that the female Mayflies live for one day only, then within twenty-four hours there would be a fall of spent gnat. As is well known, big hatches occur spasmodically during the whole of the Mayfly season and, even though a big fall of spent flies may come the following afternoon or evening, no one can say that these were hatched the previous day. In fact, they might have hatched at any time since the start of the season.

Four years earlier I had forecast that 1950 would see the Mayfly on the Upper Avon once again at the peak of their hatching and the first days of the season bore out the prophecy. Big hatches of Mayfly occur in cycles and it seemed that the succeeding five seasons should be good ones. After this, unless something is done to introduce fresh stock, a decline might follow, for Mayflies are no different from many other creatures in nature. Continual inter-breeding will wipe out living things of all kinds, more surely, if more slowly, than any device or poison yet invented or discovered.

Fortunately the introduction of a fresh strain is a feasible operation. The method is simple. I made arrangements with a keeper on a river some twenty miles away by which while the Mayfly season is at its peak, I should catch up a few hundred freshly hatched female Mayflies, put them in light cardboard boxes, and take them with me to my friend's water and there turn them loose. The next stage was for us to catch up as many females as possible from his water to be turned loose near the fishery I look after. This collection of female Mayflies is quite a simple operation. Most rivers breed the two most common species, and a watch near a place where flies are known to hatch prolifically shows that most of the insects make for a particular sheltered spot, and there take cover. All one has to do is gently to pick them off the leaves, with wings between thumb and forefinger, and drop them through a small hole in the lid of the container. They should be turned loose, a few at a time, in sheltered spots near the riverside, and then one has to hope that weather conditions will be sufficiently favourable for mating and egg-laying.

Though many people say they hate a Mayfly season, I am doubtful if at heart they really mean it. What is there to hate about it? Indeed it should bring pleasure, for no one is forced to fish during this season, and they should be happy in the fact that the trout rapidly attain an excellent condition and will provide better sport in later months; and who can see these beautiful insects about the riverside without appreciating their gracefulness and the eagerness with which the trout take them?

I would like to see big hatches of Mayflies occurring year after year without a break, as they did when I was a boy; as they did from 1927 to 1930, and on the Upper Avon and elsewhere about the country in 1950. I want to see the sunlight glinting on a hundred thousand wings as the males dance in the meadows, and see the dipping and curtseying of the females as they sow the seed for a future generation. I want to see the whole surface of the river broken by the rises of excited trout—monsters brought to the surface by the temptation only a Mayfly hatch can present, and the angler standing by me to be no less excited. For it seems to me that without the Mayfly season there is a woeful gap in the charm of a river valley during mid-May to early June.

Hatching with the Mayflies on south country streams one may see a sulphur-coloured insect which, when viewed at a distance, bears some resemblance to a Mayfly. Indeed, fishermen often mistake it for one of the ephemera species. Close examination will quickly prove it to be of another genus, for it has but two setæ—a feature quite sufficient to tell the two apart. It is a handsome and beautifully coloured insect, and of a size approximating to that of the smaller males of ephemera. This fly is *Heptagenia sulphuria*, more commonly known as the Little Yellow May Dun. Excepting the Mayflies, it is the largest of the duns to hatch on a chalk stream.

Little Yellow May Dun is a very good description of the insect, excepting that it is little only when being compared to the Mayflies. So far I have neither seen nor heard of anyone who has been successful in getting trout to take representations. Some old writers say the insects have a bitter taste, and that this is why the fish will not take the naturals or artificials. Well, it is perfectly true that trout, either small or large, will rarely take these insects, but I find it hard to know how fish can dislike the taste without trying them. Birds take them, even as they do the Mayflies.

Somehow I cannot think *Heptagenia sulphuria* is a native of the south country, for, though a few may appear seasonally in the chalk streams, there is no successful regeneration of the species. The numbers which hatch yearly alter but little, and this regardless of varying conditions. Possibly it is because of the environment that is needed by the insect while in its larval and nymphal stages, for these are inhabitants of more turbulent classes of water than usually are to be found in the south.

The larvæ and nymphs are of the clinging, or crawling, group, and in the more rapid and well aerated parts of the river live in the dark beneath stones and other debris. Hatch pools, weir holes, breakwaters and any area down-stream of a fall in the water are the places most favoured by them, and here the creatures cling like limpets to their shelters. They are very poor swimmers, but their extremely powerful and sturdy legs, combined with the flatness of their general structure, allows them to move rapidly while on the bed of the river, on stones or weeds, and they can maintain their position even in the strongest of currents. The nymph is an ugly creature, and often of a colour that, to the naked eye, is almost black. It surprises me how such a beautiful fly can emerge from it with such a contrast in coloration.

It is doubtful if trout see many of them in larval or nymphal stages, and never yet have I found one in the autopsies I have carried out throughout the year. When hatching they make use of anything which reaches from river bed to the surface, crawling up it until they reach the air. They hatch very slowly. The time elapsing in the change from nymph to dun will, at times, take as long as half an hour, and all the while the nymph is just beneath the surface of the water. Actually, the appearance of duns floating on open water is not a frequent sight and, therefore, fish have little opportunity to take them even if they wished so to do. I think this is the answer to the question as to why the Little Yellow May Dun is disregarded by trout. Hatches are spasmodic. At no time are there sufficient numbers for trout to get accustomed to seeing, and to taking, the larvæ, the nymphs, or the duns. When the occasional dun passes over the fish it is ignored, even as trout will ignore the first few Mayflies of the season.

The same can be said of the spinners of *H. sulphuria*. Trout seldom get a chance to eat them, for it is only at odd times that either male or female is to be found upon the water. Mating takes place well away from the river, at times a mile or more distant, and the female then proceeds to the river to lay her eggs upon the surface. The process of egg-laying is very much like that of *Centroptilum* (the Spurwings), just a dance above the water until all eggs are extruded to be carried on the oviducts, then one or two dips to deposit them. The eggs having been laid, the female, far from exhausted, whirls away up into the air, and is lost for ever as food for fish. She then dies in the vegetation of the surrounding countryside.

One of the questions which every river keeper has to answer many times in a season is "Do you think there will be a good evening rise?" It is not always an easy one. A good rise of trout can be influenced by several different occurrences, but in any case, whether in daytime or in evening, the rise of fish has to depend on insects borne in or on the surface of the water. It would seem that for many species of flies the cool, calm atmosphere of late evening is the best time for them to lay their eggs, and sometimes, when the weather has been exceptionally hot and the river low, some species may choose to hatch after sundown, when they find it to be easier to break through the film which, throughout the day, has proved a barrier on the surface.

There might be a combination of the two—an appearance of spinners and a hatch of duns.

But there are times when one can have a very good idea that there will be sufficient insects on the water to interest the trout. With the exception of one or two species, it is the habit of the ephemeroptera to mate just before egg-laying. It is invariably the custom of the female to go to the male when mating is desired, and so great companies of males gather about the meadows and other sheltered places to await the coming of the other sex. It would appear that these males know when conditions are favourable for egg-laying. They know also that the females are hidden in the vegetation of the surrounding countryside, and of the numbers that are likely to appear to be mated. So hour after hour, long before the females leave their shelters, the males can be seen dancing up and down in an endeavour to be in view when the females need them.

Often an evening rise has to depend entirely on a fall of spinners—a fall of female spinners which have come to the water to lay their eggs. There are occasions when many hundreds will complete their egg-laying and then fall spent to the surface of the river to die. There are many small flies which lay their eggs by dipping, but many others deposit their eggs beneath the surface of the water and after doing this will rise dead or dying to the surface there to float in or on the film which covers it. Occasionally, flies of other genera, which carry and drop a compact ball of eggs, such as the Sherry spinner, the female of the B.W.O., will fall spent to the water and die.

So, though there can be no hard and fast rule, a glance about the meadows near the riverside can give a clue to an evening rise. Large assemblies of male spinners dancing at early evening can be taken as a good indication that females are expected and that egg-laying will take place in a few hours. The quantities of males give an idea as to the number of females which may come to the water, and if females lay eggs, then it is quite certain that many of them will die, and will be present in, or on, the surface of the river, to become the playthings of the vagaries of the current. This, however, is as far as I can go. Usually, if sufficient flies are on the water, trout will rise and take them but, as we all know only too well, there can be exceptions.

From June to August the Brown and Black Silverhorns are a familiar sight about the river during warm evenings and, I should think, that to fish they are the most exasperating of all the flies they see over the water. Occasionally, the weaving hordes will goad a little trout to a point beyond endurance and in a fit of rage he will make a leap in a futile effort to catch one. Silverhorns are of the Sedge family. They are true river flies, for both species, the black and the brown, spend their aquatic life as caddis. Both are numerous in all classes of water and have habits that are similar. The brown one is the larger of the two, and both species are known as Silverhorns, owing to their long, narrow-jointed antennæ, or horns, which glisten like silver when the flies are held to the light.

In construction and appearance the caddis case bears some resemblance to the Grannom, but Silverhorns hatch from where their cases are tethered to stones and other debris on the bed of the river and, unlike the Grannom, have no preference for fast-flowing water. The fly, enclosed in its nymphal envelope, rises swiftly to the surface, to hatch in a manner not unlike the ephemeroptera. Leaving the nymphal skin floating on the water, the fly quickly scuttles away to the nearest cover. A period of a few hours may elapse, then the males congregate and start their characteristic weaving dance just above the surface of the water, so close indeed that sometimes they ruffle the surface with their wings. In either species the male is the smaller of the sexes.

It is while Silverhorns are engaged in dancing that they are mostly to be seen over the water, for they may gather in many hundreds at late afternoon and continue to dance unceasingly until darkness makes further observation impossible. As I have said, these congregations are males. If a study is made of them as they are dancing, it will be noticed that the females will come in ones and twos from the surrounding countryside and fly into the weaving assembly. I have spent many hours watching them and always with interest. A female is quickly noticed, she is seized by a male, and then, carrying her mate, she rises almost vertically from the water, and hastens to shelter in the vegetation at the riverside. A period of several minutes may pass, then the male leaves her and goes back to rejoin his sex and to continue with the dancing. The female may cling to her perch for an hour or more, then she flutters down and flits about just above the surface of the water, while making continuous dips to touch the water and lay her eggs.

And so, about mid-evening, a horde of male Silverhorns may be seen in a concentrated group, spinning around and around, this way and that, in an unceasing movement of activity, while all about the area of water nearby the females can be noticed as individuals, dipping to touch the water. The female is not so exhausted by her egg-laying that she cannot leave the river, and later will die on land.

Big trout have learned the futility of trying to catch these flies while they are weaving above the surface, but should a pair, by mischance, fall to the water they may struggle for a few moments before separating and in this brief time will form an attraction which no trout seems able to resist. A representation of a single fly is useless, for in the natural course of events it is seldom that an individual rests upon the water for more than a fraction of time—a fact undoubtedly known to the trout.

At times, I have used a representation of the paired flies with great success for trout and grayling. Though artificials tied to represent the hatching nymph can be very deadly if fished with a dragging motion just beneath the surface, I do not advocate them for trout in a chalk stream. With a pattern so tied, I once had great fun while fishing in a lake for rudd. Chub and dace will also take them readily and one can catch fish after fish from a shoal without unduly alarming the others.

Loss can be caused to a river's stock of flies through

many causes. Some are difficult to anticipate and impossible to prevent. One evening I was standing on a bridge which crosses and carries a minor road over the river. Just up-stream a set of hatches control the reach above, and down-stream is a deep pool caused by the continual scour as water rushes beneath the bridge and then falls in seething spouts to the level below. Down-stream of the pool is a wide, shallow reach where the water ripples and aerates as it passes quickly over the bright gravel bed. This shallow and pool, hatch structure and bridge is a favourite egg-laying place for many species of ephemeroptera, and I often stand near the parapet of the bridge and watch the activities of the insects as they go about their business of reproduction.

Many a time I have seen the egg-bearing Sherry Spinners come in their great processions to this bridge, and there, finding the impounded water up-stream to be unsuitable for their purpose, turn back down-stream to lay their eggs in the pool, or about the shallow just below. It was getting late—the sun had disappeared behind the high elms and beeches in the wood on the west side of the river, and dark shadows fell across the water. Though warm, the weather was not settled. About an hour previously a rainstorm had swept throughout the valley, and all around the vegetation glistened with moisture. To my left and right, the tarred road, wet with the evening rain, wound away like an uneven ribbon of polished black steel.

As I watched, a big procession of Sherry Spinners came sweeping to the bridge. Instead of turning back to the shallow and pool to lay their eggs many of them swept up, and over the parapet, and there milled around in confusion above the middle of the road. I could see they were undecided —that the junction of river and road puzzled them. The procession broke up in disorder, some went up the road to my left, others to my right, and a few went back down-stream. To me now, the road, with its wet, shiny black surface, looked like a stream, for the uneven tarred surface rippled in the evening light like water running fast over a rough river bed.

And as such it must have appeared to the Sherry Spinners, for many thousands left the river and followed the course of the road. Some dropped their balls of eggs as they flew along, but many others settled on to the road, still carrying their egg masses, and soon the road for about one hundred yards each side of the bridge was a mass of dead and dying flies. Now and then a motor car swept through the ranks of the flying insects and smashed the creatures into pulp on wings, bonnet and windscreen, while the tiny balls of eggs tumbled to the side of the roadway. In less than half an hour many thousands of flies died on that short stretch of road, and many thousands of eggs were wasted. Long before morning the road was as dry as a bone—the following day, the hot sun withered the wings and bodies of the dead flies, and dried up the balls of eggs, so that they became the playthings of the varying air currents.

Though I made a note of this incident, I have seen the same tragedy occur many times. The progress of Nature has not kept in tune with the progress of man, and it is unlikely that it ever will. In the past we have gone to great trouble to see that our roads are dressed with a substance not likely to be harmful to insect life should washings from them enter a river or stream, yet, so far, little thought has been given to the fact, that, during inclement weather many millions of insects are each year being deceived into thinking that our roads are waterways, and not only waterways, but ideal places for the reproduction of their species.

Such a wet uneven surface of a tarred road is like a rippling shallow of a river—it has the same mirror-like effect in reflecting the sky and surrounding vegetation— and while the roads of our countryside remain as they are at the present time, the river insects will continually be deceived. Unfortunately, the Sherry Spinners are not the only flies to make mistakes. I have seen the same thing happen with Mayflies, I have seen Olives crawling down the tyres of stationary cars into puddles, and I have seen them settle on my boots and crawl beneath the sole. To me this waste of life is indeed a tragedy, for I feel sure that if road surfaces in the immediate vicinity of trout streams were made of a different colour, all this waste could be avoided.

8 DECEPTION

THE art of dry-fly fishing lies in the ability of an angler to present an artificial to a fish in a manner which will deceive the fish into thinking the offering is a natural insect. No doubt, since the art of fishing with an insect as a bait was first practised on river and stream, a study of the flies of most value has been carried out. For a while the natural insect was used on a hook exclusively and then, as time went on, representations were made, and the artificial came into use. In the years that have followed it would seem that almost every insect of river, and riverside, has at some time been imitated. Natural flies have been taken from the water and from the surrounding countryside, and in many cases painstaking copies have been made.

Often as many as a score or more tyings have been made, which show an insect in its various stages; tyings which imitate the fly as it is hatching; as a dun, and as a spinner. One fisherman once told me a Red Quill was a representation of a female Medium Olive after it had been hatched two days. Well, perhaps it is, but what I would like to know is how a trout can (in the short time he has to study it) know whether the fly has been hatched two days, or a week, for it needs a microscope and a keen eye to detect the difference once the fly has changed to a spinner. If it has not changed to a spinner it is very unlikely to be found on the water.

However, this one instance is sufficient to show the meticulous care taken in years gone by to try to present to the fish the kind of fly for which it is looking. In the old days there was a school of fly-dressers known for its devotion to exact representation. Great care was taken to copy every detail of the insect both in colour, size and sex, but though detailed accounts of the materials used for this purpose and method of dressing were thoroughly explained the most important part of all was completely neglected.

I am in favour of an exact representation for dry-fly or nymph, but when I use the word representation I try to carry it into effect. I want representation that is satisfactory from the trout's point of view. A correct imitation of the colouring and shape of an insect is not enough. To represent truly an insect, and successfully deceive a trout, the artificial must be offered at a time when he is taking, or likely to be taking, the natural from or beneath the surface of the water; offered in such a manner that it looks alive, or dead, or in a semi-inert stage, as natural ones are at the moment.

One has to take into consideration that a fly dressed with red hackles may look green from the trout's point of view against a blue sky, and that on a dull day a light fly may look dark. Waves may distort the size of a natural insect, and make it look to be twice as big, while the calm water of eventide may show every detail. A certain insect may struggle desperately while on the water, while another, though alive, will float without movement. An artificial fly which looks perfect as it lies in the palm of your hand, or as it is held up to the light, may, though it conforms to the textbooks of fly dressing, be a disappointment. By contrast, a roughly tied fly with a happy combination of colour and size may look exactly the thing for which the trout are waiting.

Since nymph fishing became a practice on many of our rivers and streams, much has been written about the speed with which trout or grayling will eject an artificial. Some writers put forward the theory that most of the nymph patterns that are made as actual representations of the natural have body materials that are too hard, and that the fish quickly discover the deception when they hold such imitations in their mouths. But it seems to me that, while we have to put the dressings of our nymphs on to a steel hook, this question of hardness is never likely to be eliminated. It would seem that some anglers imagine that the jaws of a fish must invariably close across the body of a nymph. Do they? What happens if a body is made of a soft substance and the jaw pressure is directed on to the bend of the hook, or if the artificial is gripped with pressure at head and tail. A nymph is not taken by a trout in the same manner that a pike seizes his prey. These tiny creatures may make an entry at any angle, and the hard substance of the steel is bound to be felt in an artificial, no matter what dressing has been used to conceal it.

Candidly, I do not think hardness or softness of body dressing makes the slightest difference. Neither do I think a fish discovers the deceit by his sense of feeling. A trout does not immediately eject its natural food and much of this is of a nature that is quite as hard as any of the dressings that have been used—or are likely to be used—for artificial nymphs. He will take beetles, water boatmen, shrimps, snails, and various other creatures that are hard and, after all, the covering of nymphs is not likely to compress with the slightest pressure.

The fact, I think, is that fish discover the deceit, not by their sense of feeling, but by their sense of taste. How quickly we ourselves will spit out say a maggoty part of an apple or, as happened to me once, a lump of soda which I took from a shop counter in mistake for a cube of sugar. The action with us is impulsive and spontaneous. We sense immediately through our taste organs that something is wrong, and out it goes. It is the same with fish.

Many fishermen must have found an artificial that has already killed a fish is often held longer when a second one takes and that a second fish shortly after the first is much easier to hook. This is explained by the fact that in hooking, playing and killing the first victim a certain amount of fishy flavour has been imparted to the dressing, to the hook and to the cast. When a second takes soon afterwards the artificial has a taste something similar to what he is expecting, and he is less suspicious of deception. So here is a tip for nymph fishermen. A nymph pattern is more readily taken and held by a trout if it has first been well covered and soaked with the slime of fish. The same slime wiped on the cast will make it sink far more quickly than glycerine.

Trout will eject an artificial nymph that is dragging much more quickly than one that is drifting loosely, so to speak. In these cases it matters little how your nymph is constructed, or with what it is anointed. The deceit is then discovered by the sense of feeling, either by the prick of the hook point as the dragging weight of cast and line moves the artificial in the mouth of the fish, or by sensation as the cast touches a part of the tongue or jaw. I think that grayling will discover the deceit much more quickly than trout, but after getting the first one and treating nymph or lure with its slime I have often killed up to thirty fish from the same pool and have had little difficulty in hooking them. One afternoon I had killed about half a dozen grayling and had then made a cast to a point that was much deeper. I failed to notice that something had taken the lure and, on lifting out the rod, thought I had hooked some vegetation on the bottom. I lifted sharply to disengage and then found I was into a fish. This fish proved to be the heaviest roach I have ever killed on a rod and it weighed 2 oz. short of 3 lb. To me it seemed that the roach had indeed appreciated the fishy flavour of the artificial shrimp I was using, and was in no hurry to eject it.

During this latter part of the season there are times throughout the day when clarity of water, combined with clarity of air, allows one to see every movement of a fish even though the water may be three or four feet deep. Under these conditions I get great delight in deceiving a trout and then watching its reaction when discovering it has taken something unpleasant, and unpalatable, for at this time I often try out new patterns of nymphs I have evolved. As I have no desire to catch the fish I use the very finest of casts, for without the necessity to hook or play the trout, all I require is a cast just strong enough to enable me to throw the nymph. In this way I have found out just what happens when a fish takes an artificial nymph, and have discovered how quickly it finds out it has been deceived, and the speed with which the offending particle can be ejected.

Nymph fishing, when carried out as it should be, is, I think, the most fascinating of all kinds of angling. Each season I see more fishermen are practising the art, but in many cases I find that, in their inability to know just when a trout has taken the artificial into his mouth, and closed his jaws, or in allowing the fish just that split second too long to discover the deception and eject the hook, far more trout are scared than caught.

Good eyesight is a great help in nymph fishing, indeed I think it is essential, for in this class of angling the indication of a take by a trout is often so slight, that it can be missed by the most keen observer. I tie my nymph patterns so that they have a quick entry to the water. The fine cast offers little resistance as the fly sinks towards the bottom, and usually the movement by the nymph as it is sinking is quite sufficient to attract the attention of a trout. It is only natural that he should be attracted, for it is the habit of many kinds of nymphs to play about between surface and river bed. This sinking slowly towards the bottom is one of their characteristics, and trout know it.

We will imagine that an artificial has been cast success-fully to a trout lying in an up-stream position, say, a foot beneath the surface. The nymph has entered the water with very slight disturbance about two feet up-stream of his head, and is now sinking and drifting slowly towards the fish. We have a clear view of the trout and as we watch intently we see his reaction as the artificial enters his zone of vision. The moving nymph at once attracts his eyes, a tenseness passes through his body, and then tail and fins synchronise in a movement to propel him forwards to intercept it. Without the slightest suspicion all fins check in outspread movement and the trout, for that moment stationary, opens his jaws to take the deception, and at once closes them again.

In this short time many things happen; the nymph is now resting on the tongue of the fish, gripped tightly between the tongue and the roof of the mouth, while the fish opens its gills to allow the water he has also taken to pass out. But while this is happening we have noted the movement of the cast and line which should be floating on the surface. As the trout took the nymph the cast stopped its drift down-stream, just a momentary check, and then the tension, as the fish holds the nymph between tongue and roof of mouth, causes the cast to draw slightly beneath the surface. It is but a fraction of time, but sufficiently long for a trout to discover he has been deluded into taking some alien substance, and he quickly makes a move to eject it. He has felt the drag of line and cast at one side of his mouth and is scared. His jaws quickly open, and with a shake of the head, the artificial is, in most cases, successfully expelled. As he shakes his head, so a well-defined jerk is given to the cast and line floating on the surface, a movement so often mistaken by fishermen to be the trout taking the nymph instead of releasing it. To be successful in hooking trout, the line should be tightened during the short period the artificial is held on the fish's tongue.

For some time I have been of the opinion that the hatching of ephemeroptera flies depends to a great extent on the condition of the surface film of water, and I quote this experience as evidence in support.

About mid-morning on 2nd October 1948, I was studying a reach of water immediately above a set of hatches; all the hatches were closed and the water was running very slowly towards an overflow. As I watched I saw a number of different flies—Olives, Pale Wateries, Lesser Spurwings and Iron Blues hatch and fly away, while many more were floating gradually down-stream. Trout were up in the water, and were on the feed. Some were taking duns, but most of them were swinging from side to side, intercepting nymphs. Apart from flannel weed on the bed, there was no vegetation in the river at this point and the water was crystal clear.

I sat near the edge of the river to watch the hatch of fly and the rising trout, and as I studied the water I saw a number of nymphs leave the flannel-weed bed and swim rapidly to the surface, but here, instead of breaking through the film and hatching or returning to the bottom as is often their procedure, the insects swam quickly along just under the surface film. The surface was perfectly clear and as each insect swam along, its small body caused a disturbance, a tiny ripple, a movement quite easy to follow. Though I watched several of the nymphs swim a distance of six feet, the general swimming journey was not more than two or three feet. Each time the insects stopped they tried to force their thorax up into the air, in which some succeeded quickly, but most of them made anything from six to a dozen attempts. Immediately the thorax came through the film, the shuck split and the fly emerged—in most cases to fly away almost at once.

Now I was rather interested to know just how far the nymphs would float or swim down-stream before hatching, so I carefully watched four different insects, two Olives, a Pale Watery and an Iron Blue. To do this, I went up-stream about eighty yards above the hatches, and as soon as I saw a nymph reach the surface film from the bed of the river I followed it as it moved with the slow current down-stream. The first, a Medium Olive, hatched after I had taken thirty-seven paces; the second, a Pale Watery, when I had taken but seven. The third and fourth drifted down together, the Iron Blue hatching when I had taken sixty-one strides, but the Olive (another Medium Olive) failed to hatch until it reached some floating weeds which were checked above the set of hatches. I saw this nymph cling tightly to a strand of weed and then come crawling through the film of water to hatch. On their journeys down-stream many attempts were made by each nymph in an endeavour to hatch, swimming here and there, struggling to and fro, and at times drifting completely inert.

All this interested me greatly. Here was a big movement of nymphs, and a hatch of flies, to be studied as easily as in a glass aquarium, flies of four different families, all hatching together, and all finding the same difficulty. As I watched the water I could see that the surface was moving in three layers, three different films, and the toughest of these was that nearest to the air.

I lay down at full length with my head over the water and watched intently, and cursed myself for not having a magnifying glass with me. I could see several nymphs playing about on the river bed and then one came gliding up towards my nose. I was looking into a kind of back water behind a jutting bank, and my eyes were not more than six inches from the nymph as it reached the surface. I saw it struggle to force its way through the film, and the film bulge upwards as though made of elastic. Again and again it tried, with pauses to regain energy. I counted the number of attempts and on the seventeenth it succeeded. The thorax came pushing through to the air and immediately split just in front of the wing cases, and without any difficulty the fly, a Medium Olive, emerged and took flight. I watched others, Pale Wateries, Iron Blues and Lesser Spurwings, all hatch at the same spot and all had the same difficulty.

Pondering over this, I moved once more down-stream to the hatches. Above these was a collection of cut weeds and other debris packed tightly together and covering some forty square yards of water. Almost in the centre of the mass was a tiny pocket, a space about eighteen inches in diameter where the packed weeds were submerged, leaving a little pool of water about six inches deep. The surface of this little pool was covered with a sort of scummy film. This film was being continually agitated by some movements below it, and for a moment I thought that a shoal of tiny minnows were trapped in the pocket, but as I watched closely I could see that I was mistaken. The agitation was caused by at least fifty nymphs, which, after my previous experience, I could see were trying to hatch. I sat down and watched closely for ten minutes or more without taking my eyes from the spot, yet not a fly hatched. The agitation of the surface became more and more pronounced.

I could see some nymphs struggling and swimming about, and others on the weeds, while some were just floating inertly beneath the scummy film. Now here, I thought, is a great chance to try out this theory of mine, and I looked around for a long stick. I failed to find one, but lying nearby were several small stones. I tossed one of these into the centre of the pocket and it had the effect I desired; the surface scum was immediately dispersed, leaving about four square inches of clear water. In a moment three flies hatched in the cleared space, and shortly afterwards two more; then the scum formed over once more. I tossed in other stones and completely dispersed the film of scum, and in five minutes twenty-seven flies hatched from that little pool; amongst them I recognised Olives, Pale Wateries, Lesser Spurwings and Iron Blues.

Thinking it over, I came to the conclusion that not only did the stones break up the surface film by splashing it away on all sides, but the action also aerated the water to a considerable extent. A thick film of water is generally present during the late summer and autumn, and I think is caused by gases rising from the rotting vegetation. It is not so noticeable when a river is flowing along at a fair speed, but at that time the Upper Avon had been exceptionally low and the temperature much higher than is usual for October.

One learns a lot by hatching nymphs artificially, but before nymphs can be hatched for identification one must first obtain them from the river, and separate them into what possibly is the correct species and sex. It is rather a delicate business, for to ensure good hatching results the insects must be handled with care and treated so that no part of their bodies, wing cases, legs or setæ, is crushed or broken. To save keeping the nymphs longer than necessary in the hatchery it is much better to catch them at a time when they are within a few hours, or days, of changing into duns.

To catch up nymphs so far I have found nothing better or easier than to make a net from a lady's stocking. All that is necessary is to cut off the foot and tie a knot in the end. The top is then attached to a framing of $\frac{1}{4}$ in. wire, made something like the frame and handle of a squash racket. The fine mesh of the stocking will hold insects of very small size. Many of the nymphs of the small flies, such as those of Olives, Iron Blues, Spurwings, Pale Wateries and Cloeon, live amongst the vegetation. In different classes of water they may be obtained at almost any time, though perhaps in varying stages. Blue Winged Olives and Caenis live mostly on or near the river bed, and can only be caught in numbers when they crawl up weed stems to hatch.

To anyone who has knowledge of the flies of the ephemeroptera and of the seasons when they may be expected to appear, the finding and catching of mature nymphs of a particular species is greatly simplified. Most nymphs of the small flies mentioned have a habit of crawling to the surface on the vegetation, there to remain just beneath the water for several hours, perhaps a day or even longer, before hatching to a dun.

When I want mature nymphs I take my net and wade into the weed-beds on the shallows. The method I adopt is to push the net beneath weeds, lift them up once or twice in a shaking movement to dislodge any insects, so that they wash into the bag of the net. This I repeat time after time throughout a weed-bed until reasonably sure of a good haul. Where water is sluggish and deep it might be necessary to attach the net to a long handle and sweep into the weeds within reach. This latter is often needed in searching for Spurwings and Cloeon.

On the bank nearby, I have a large white enamelled bowl, half a dozen 1 lb. size jam jars, and a spoon with a long handle. These, together with a pocket lens, complete my outfit for sorting out the various nymphs captured. About an inch of water is needed in the bowl and then the contents of the net can be tipped into it by turning the stocking inside out. Nymphs and larvæ of several species may be present in this one haul, and all show up plainly against the white background. Soon it is possible to be sure of different kinds and then the spoon can be brought into use. I use a spoon because this is less likely to cause damage. When a nymph is located it is quite easy to encourage it to enter the spoon and so be lifted from the water. A moment or two under the lens makes certain of sex and characteristics, and then it can be transferred to water in one of the jam jars.

Males and females can be determined by the size of the eyes, and one need go no further. The eyes of the male are the larger. I try to sort them all out, keeping the sexes separate. Each species has some characteristic either in shape or colour, in movement or in size, by which they can be separated, but the job needs good eyesight, a steady hand and lots, oh, lots of patience.

Nymphs of the flies I have mentioned may be obtained throughout the season from April to October by search in the weed-beds. To get them all, a search once a fortnight should be made. Mature nymphs of the Mayflies are much easier to collect. These live in the river bed and in early May can be found almost everywhere in the gravel at a depth from 1 in. to 18 in. All that is necessary is a garden fork. Dig up a section of the river bed where it is clean and firm and then examine the spoil. The nymphs will come crawling out of their tunnels and caves and are easy to pick up between your thumb and forefinger, and can be dropped into a container.

April and early May is a good time to find mature nymphs of such flies as the Yellow May Dun, Turkey Brown, Claret Dun and *Habrophlebia fusca* which for the want of a name other than its Latin one I call the Three-tailed Iron Blue. Search for all these should be made beneath big stones in rapid water. Find a big stone that is being swept with a strong current and lift it clear of the water as carefully as possible. Then examine the part that has been resting on the river bed. A close scrutiny is necessary and then you may find one or more of the species. These are flat nymphs and as they cling fast they are very difficult to detect amongst the algæ or other growths which may be on the stone. Usually they are very dark in colour. I find a spoon is very useful to transfer these from the stone to the collecting jars. Mature nymphs of all species show well defined wing cases and the lens will reveal that the whole skin is loose. To the naked eye the creatures appear to have a silvery sheen over the whole body, and are easy to sort out from others that are in a stage that is less advanced.

9   THE HUMAN ELEMENT

Iᴛ is the human element that is often blamed for failure in all sorts of enterprises—sometimes justly, sometimes unjustly. In the matter of catching trout the part played by the fisherman is apt to become distorted in several ways. But most of us are familiar with this sort of thing: "Like thistledown tied to a strand of spider's web the fly touched the water and, with hackles turning and sparkling in the sunlight, it drifted slowly towards the waiting trout."

Somewhere, I cannot recollect where, I read that passage. Perhaps it was someone's poetic version of the casting of a dry-fly, but perhaps the writer really meant what he had written and that was the manner in which he would like to see an artificial fly presented to a trout. If he did, I agree with him wholeheartedly, for though I have seen many thousands of flies cast to rising fish, I fear most of them have fallen (somewhat) short of this ideal of perfection.

Many fly fishermen use oil to make their flies float well and to sit up in a lifelike manner on the water, but I never recommend the use of such aids when I teach others to cast a fly. If flies are dressed sparsely and on light hooks, used with a cast that in its fineness is in comparison with the hook size and dressing and then cast in a proper manner, those flies should float, and not only float but to the fish look sufficiently lifelike to deceive them into thinking they are live creatures instead of bundles of fur and feathers.

The words "turning and sparkling in the sunlight" interest me considerably, for unless a fly is cast so that it falls like thistledown the hackles cannot possibly turn and sparkle in any light, for most of the fibres will have penetrated the surface film and are there held fast in its grip. No matter if a fly has been oiled, or treated in some other way to make it float, if it falls heavily to the water it breaks through the thin surface film and immediately becomes a thing that is dead.

Stiff, sparkling hackles are essential with a dry-fly, and these should be spun around the hook so that they act as a support to keep the body cocked above the surface. Drop a well-tied dry-fly on to a glass of water and watch carefully the reaction from below, as the moisture begins to soak into the hackle fibres. You can then see what is meant by "turning and sparkling in the sunlight," for, as the fly settles, so the hackles contract and move. The whole artificial becomes a thing of movement and apparently of life. I feel sure it is movement that is the greatest attraction to trout. One memorable day I was watching a trout that was rising steadily close to me. More than a dozen times the fish came to the surface for particles of vegetable matter that were given movement by the vagaries of the current in which he was poised. Some of these he took into his mouth and, though he instantly rejected them, this does not alter the fact that he had been deceived, and, had either of these particles contained a hook attached to line and rod, the fish could have been hooked and perhaps landed.

With nymph fishing, as with wet-fly, one can, if necessary,

impart a lifelike action to the artificial by movement of the rod, in any case the nymph or wet-fly has, without any movement of the rod, a certain attractive motion as it sinks towards the bottom. But with dry-fly, any drag of fly or cast usually scares the fish. The artificial must therefore be constructed and presented in such a manner that it has lifelike attraction without other assistance.

Every dry-fly fisherman must have experienced days when even the shadow caused by a falling line and fly is sufficient to scare a trout, especially in water where cover is scarce. We like to assume that the fish are educated and know we fish for them, but though I know, and do not under-estimate, the instinct of trout I hardly think they have sense enough to reason things out so clearly. A shadow to them, no matter what the pattern, is something to be feared. One sunny morning I had cast a nymph to trout of about $1\frac{1}{4}$ lb. He was feeding well and I knew I stood a very good chance of getting him. When he moved forward, I was sure he had sighted the nymph and intended to take it, but at that very moment a rook passed between the fish and the sun, and the shadow of the rook fell on the water just in front of the trout's head. The shadow passed quickly, but the trout moved even faster and dived headlong into a bed of weeds, five yards up-stream.

Now this was not the first time by many that I had seen trout scared by shadows—not always of rooks, but of many birds—and often I have seen the shadow cast by an aeroplane travel up and scare every trout in as much as a quarter mile of water. But I was in a meditative mood. I began to think about this shadow effect, and to wonder about it. Times without number my own shadow has put down a rising trout, scared a school of grayling or disturbed a basking pike. Again I have seen shoals of roach and dace scatter in all directions. I came to the conclusion that the alarm must be caused by some inherent instinct still prevailing after hundreds of years, in fact, that a shadow stirs quickly a memory of winged terrors of a bye-gone age. To fish, a shadow means an enemy from the sky. Even to-day there are attackers which approach fish from the air but on the Upper Avon these consist of only the heron and the king-fisher, and no great numbers of either. But perhaps many years ago this trout stream was the hunting ground of other winged predatory creatures which are now extinct; perhaps also it was the haunt of ospreys, eagles, cormorants, gulls and the like, birds which even to-day are a source of terror to the fish of the locality in which they hunt.

I have many times seen kingfishers at work, hovering a few feet above the water after a speedy dash up-stream and the quick plunge they make to secure prey, and I have also had some good studies of the heron in action. Though perhaps the habits of the kingfisher have not varied through-out the ages, I think those of the heron might have changed. I occasionally see the use of a habit which at one time might have been a chief one. We look upon the heron as a wader, indeed he is formed for this purpose and most of his prey is killed while he is wading, especially trout, but herons do take fish from deep water while in flight though it would hardly be correct to say they hover. The method they adopt is to fly up-stream just clear of the water and to drop suddenly as the prey is sighted. Legs and beak strike the water together and with a downward thrust of the feet the bird lifts himself clear of the water with, if he has been successful (as in many cases he is), the fish impaled on his beak. I once saw a heron spear a large fish in this way, and apparently it was too heavy for him to lift. He failed to thrust himself clear of the water. But he was not in any way alarmed and though he lost his fish he swam to the bank without trouble, and was immediately able to take flight. Herons do hunt a lot at night and I think it possible that they then attack from the air more frequently, and if this is so, then trout have very good reason to be scared of shadows.

Throughout the years I have been a river keeper I have seen many hundreds of trout take an artificial dry-fly only to be missed by the angler in his attempt to hook them. It is indeed rare for a trout to rise twice to the same pattern, or, in a short while, to take another artificial in the same confident spirit which he showed in taking the first. And so, throughout the season, opportunities to catch fish are continually being lost, fish are being educated, and in consequence many anglers are disappointed in the result of their fishing.

There are days on a chalk stream when rising fish are few and very far between, when, if one is to make a bag, every opportunity that is presented must be made to count. The correct tightening of a line at the critical time can often make all the difference between a blank day and a return with a well-filled basket. I have many times spent whole days with fishermen, seeing them rise and miss fish after fish, and put down many others they might have risen had they exercised a little more control over their emotions. In many cases, these fishermen are prone to blame everything but themselves for their lack of sport. Some of these have been very skilful in the presentation of their flies and have been successful in deceiving trout into taking their offers and yet have failed to appreciate the correct moment when they should drive home the hook.

The successful hooking of trout with a dry-fly must be a cool, calculated, precision job, a perfect unity of hand and eye, for in many instances the whole thing must be carried out in but a fraction of time, and a mastery of emotions during this moment makes all the difference. I compare the hooking of trout with a dry-fly to target shooting with a miniature rifle, for in rifle shooting the same synchronisation of hand and eye is necessary to ensure hitting the ten ring with every shot. Hesitation is fatal. There is just that moment when you know the bullet will go home if it is released, and that is the time to do it. Conditions vary considerably with both rifle shooting and the hooking of fish with a dry-fly. There can never be a hard and fast rule laid down for either. The success of both the rifle shot and the fisherman depends entirely on his own ability to appreciate the exact moment for action and the conditions which make it possible.

In rifle shooting, different lights, wind and general atmospherics all play a big part. With fly fishing there are all these and other factors with which to contend. Trout rise in a great variety of positions, in fact, it is rare indeed to find two fish on any one day of similar size that are rising in an identical manner or in places that are exactly alike.

Many times I have been asked how I know when to strike —or tighten—on a fish. It is not an easy question to answer. One cannot say pause for this or that length of time, or tell people to be governed by the size of the fish, by the strength of the current, or nature of the water in which they are fishing, for all this is much too complicated to be easily understood. There is only one true answer. The line must be tightened while the fish has the artificial held in its closed mouth. To this they immediately ask: "How am I to know?"

Well there is only one way to know, and that is to *see* the jaws of the fish close on to the fly, and when one has acquired the art of doing this, then the rest is simple. No matter where or how a fish is rising, or what fly he is taking, if that fly is floating on the water then the jaws of the fish must break through the surface to take it. Most fishermen, I think, study their flies instead of watching their fish or (if they cannot see the fish) concentrating on the spot where they expect such fish to rise and take the fly. They are so intent on looking at the fly that they fail to see the nose of the fish as it pokes through the surface. All they see is the disturbance of the water after the jaws have disappeared carrying with them the fly. A split second is lost and with it, in many cases, the opportunity to hook the fish.

No matter where or what the nature of the water, a fly whether natural or artificial, is held tightly in a trout's mouth the moment his jaws disappear beneath the surface, and if the line is then tightened, the hook stands a very good chance of taking a hold. Many times I have seen good fishermen make a perfect stalk to a rising fish, cast a fly that has deceived, wait patiently until the exact moment before tightening to drive home the hook, and then jump up into full view of the victim to try to land him. Indeed some will run up-stream or down in an endeavour to recover line or to ease pressure on the rod, and it would seem, that in a moment, they forget they are dealing with a wild creature.

Many scores of good fish are lost each season while being played out and, in many of the incidents I have witnessed, the fault has been that the fish has been unduly frightened by the angler. Personally, I think it is the duty of fishermen to do their utmost to land a fish as quickly as possible, once it has been hooked. I hate to see people dilly-dallying about and allowing a fish to wander here and there almost at will, or to scare it to such an extent that the playing is difficult. And I hate to see a good fish escape through carelessness. The law of fishing should be, take command and keep command. This is less difficult than it sounds, for to-day our rods and equipment are very capable of dealing with the fish for which they were designed.

But one thing the fisherman must keep uppermost in his mind. It matters not one jot the species or whether they are wild bred or from artificial stock. The moment a fish finds himself tethered, he fights for liberty. At the moment of being hooked he is puzzled and has no knowledge of the source from which the danger is threatened. To him then, one side of a river is no more to be feared than the other. And if you can keep him puzzled so much the better, for you have a much better chance of landing him quickly.

It pays to keep hidden or at least to remain motionless after a fish has been hooked, and then to use the tension of the rod and make the fish tire himself. Play him hard and bring him to the surface as soon as possible and make him open his mouth to swallow both air and water, for the sooner he starts to gasp the quicker he can be landed. By keeping hidden or perfectly still, one can often get the victim to come within easy distance for landing, for though a fish has certain powers of thought he is easily deceived. An angler crouched low on the river bank can be mistaken by him for one of the things he sees in everyday life. The angler has some resemblance to a tree stump, a clump of willow herb, rushes or other riverside vegetation and this strikes no chord of fear. Though puzzled, frightened, and possibly hurt, the fish feels he can pass quite close, or beneath, this inanimate object, and in doing so offers an opportunity to the fisherman to make an end.

But should this clump of vegetation suddenly assume double proportions and start to move rapidly either upstream or down, then puzzlement changes to terror, for the illusion is quickly dispelled. The fish at once realises that the thing restricting his movements is in some way connected with a living creature on the bank, and then he does his utmost to escape from both. He fights to the limit of his strength and endeavours to keep as far distant as possible from the enemy that is obvious to him.

There are, however, some very important points to keep in mind when making a quick landing. Use a big landing net, a sharp gaff, or a fool-proof tailer, and then make quite certain with the first attempt. I find it pays to make up your mind where you are going to land your fish before it is hooked, and to have your landing gear ready to slide into the water the moment the fish is on. Some may say that the sport of fishing is in playing out the fish, but with this I cannot agree. I feel one gets the peak of satisfaction when the hook is driven home and one knows deception has been achieved. To me, the playing out of a fish is but secondary, and though perhaps one may get thrills, they are mostly thrills of apprehension lest the fish should escape.

How often I hear the story of fish being hooked with Mayflies, of trout being on for but a moment before escaping, of those, to use a well-known phrase, "which have come unstuck." Every Mayfly season far more trout are frightened than caught, and I feel sure many of these are so scared that it may be weeks, or perhaps not at all, throughout the rest of the trout season before they will again rise to the surface for food. Opportunities are missed time and again, and many fish which might have been included in the annual tally live on, but are perhaps lost to the fishery as far as sport is concerned. I hear also the theories to account for this:

"I am just pricking them—the fish are coming short—are not taking properly—are not hungry—are turning off at the last moment"—and so on. Well, there never can be a hard and fast rule, but to me it seems possible that if the fish has risen to take the fly, and the angler has felt resistance when lifting his rod to tighten, then the trout has indeed taken the artificial into his mouth. The hook actually has been inside the jaws and all but one of the theories that are advanced can be disregarded.

When fishermen speak of pricking their fish, for once they are stating an actual fact. The fish are being pricked— just touched with the point of the hook in a place where pressure is needed to drive home the barb. Most of the fish which "come unstuck" are those which turn head first towards the rod immediately they feel the hook point, for the brief moment when pressure is relaxed is sufficient for a shake of the head to throw the artificial clear of the mouth. One must remember that to hook properly the artificial has to slide inside the mouth of the fish. The pressure from the rod must be sufficient to have the effect of dragging the fly from between the jaws. There is a moment when the big artificial is gripped tightly by the fish and when a strong jerk is necessary to move it far enough for the hook to go home.

Many fishermen recommend the use of stiff cocks' hackles on flies. The fly sits up better on the water they say, and it is not so likely to get caught up in the vegetation of river or riverside. Well, it is perfectly true that a stiff-hackled fly does not hitch into anything with which it might come in contact. Yet that same fly often is expected to hitch up and make a connection with the jaws of a trout. Those stiff hackles which guard the hook point so successfully as it is passing through sedges and grasses, through bushes and tree branches, can also guard the point when the fly is in the mouth of a fish.

The bigger the post the harder the blow that is needed to drive it into the ground—a tack hammer will not do the work of a sledge. If you use a big hook then you should also use the tackle to drive it. Many fishermen use the same rod, and rod top, throughout the fishing season, and they exert no more power in tightening on a fish that has taken an artificial Mayfly on a No. 6* hook than they do when a fish has taken a representation of a lesser Spurwing or a Black Midge on a OOO hook*. The result is that in many cases the hook of the Mayfly does not penetrate deeply enough to get a hold. There comes that brief moment of feeling the fish, then the horrible slackness of the line.

As every dry-fly fisherman knows, there are times during the summer days and evenings when trout may be rising well, times when, no matter what pattern of fly is used, or how well it is constructed, though every effort is made to cast accurately and delicately, the fish are not sufficiently interested to rise and take. Often fish after fish will go down after the first cast has been made, and finally, in desperation, the fisherman changes over from dry-fly to nymph fishing.

It is a problem, but often a study of the surface of the water can provide an answer, or at least give some enlightenment, for this uncanny discernment of deceit by trout generally occurs when there is a bloom on the water and to all appearances the surface has the look of being dusty. A few moments of observation will prove that there are two distinct films and that the lower of these is travelling at a speed which often is three times greater than the upper one. In fact, there are two layers, and the top one of these is composed mostly of fine particles of dust, of vegetation and other minute fragments that have been cast on to the water from the surrounding countryside. Intermingled with all this will be tiny animal life from the river, diatoms, protoplasm, etc., some dead, some dying and others drifting in the warm layer of water while in the process of reproduction.

The point I wish to make is that this upper layer, or film, is of a density and of a strength sufficient to carry a dry-fly and cast, provided these have been presented delicately. The hackle fibres rest on the film and the fly rides high above the surface. To all intents and purposes, a perfect

*No. 6 is now No. 9 and OOO is No. 17.

presentation has been made. How then can there be anything to scare the fish, or to give him reason to think he is
being deceived? There would be nothing but for the line.
Lines are made so that a fly can be cast accurately and to a
great distance, but they are heavy, much heavier than the
gut and the fly. The tendency, no matter how well greased,
is for the line to cut cleanly through the upper layer of
surface films. What then happens is that the line is gripped
by the quicker lower current and made to travel at almost
its speed. The result is that the gut cast and fly are dragged
along through the upper film to cut a lane in the flotsam,
and with his view from below this lane, as it is silhouetted
against the sky, this immediately becomes apparent to the
trout. It seems rather ridiculous to suggest drag when there
is little or no appreciable current, or when a cast has been
made up-stream and with a slack line, but there is no other
way in which it can be described. To the trout the artificial
is moving in an unnatural manner and he has every reason
to be scared.

The top layer, or film, of a river, or for that matter any
water, is the first to react to changes of temperature. The
bloom I have described is usually obvious during the
afternoons of very warm days. Hot weather quickly kills
the oxygen content, and where water and air meet there is
created a thin layer that is impregnated and linked together
with the creatures and flotsam to form a mass that is capable
of considerable buoyancy. It is as though a thin sheet of
material is spread over the surface of the water—a sheet
which successfully prevents penetration by anything but
heavy objects.

What we have, in fact, is a false surface which, when
broken by cast or line, must appear to the fish even as a
crack in a big plate glass window would to us. These layers
remain constant until the approach of the cool atmosphere
of evening and then they gradually disperse. A change in
the weather bringing a sudden storm will also break up the
film, the surface is once again oxygenated, and the tiny
animal life retreat towards the bed of the stream.

Throughout the season good rises of trout often coincide with wet days which follow a week or two of hot weather. Such days may be of wind and rain, boisterous and cold, a complete change. Yet at these times flies may hatch in thousands, for it would seem that the aerating effect of wind and raindrops provides conditions that are very favourable for nymphs of the ephemeroptera to change into duns. Usually, when there is a good hatch of flies, there is also a rise of trout. In fact, in natural conditions, a rise of fish must depend on the appearance of insects on, or just beneath, the surface of the water. Sometimes this happens in weather that is fair and fine, but I think there is something different on a wet day. Perhaps it is because the river has become alive after the lethargy of past weeks, for movement is everywhere, and there is an added interest to keep one's thoughts from straying away in search of other subjects.

These are days of hatching nymphs and duns, days when opportunities are presented to see something of the stock of insect life and the trout a water is carrying. I am often disappointed to see trout rising everywhere and yet not a rod anywhere on the riverside for, strangely enough, few fishermen take advantage of these wet days. Somehow, I think most fly fishermen are of the opinion that perfect weather is needed for hatches of fly. Many, I know, get such great enjoyment from everything about the valley that, to them, fishing, or catching fish, is but an excuse to be at the riverside. And they know a stream is only at its best when it is bathed in sunshine and shadow—when all the wild creatures of the valley are about their business of life.

Progress along a river bank when vegetation sweeps into your face or showers you with raindrops is never very nice, but I would like to see more fishermen out on these days of fly and fish, as I feel sure that, despite the conditions, they would enjoy themselves and, in the excitement of fishing, forget all discomforts. They are days which may well be remembered for the hatching of duns. Probably, because of the damp atmosphere, the insects take much longer in drying their wings. In consequence they float on the water for a greater period before taking flight. Plenty of opportunity is thus afforded for trout to see them and to rise and take them from the surface. The fisherman has no difficulty in seeing which species of fly a fish may be taking and, if he wishes to do so, no lack of opportunity in catching an insect to identify. Often, on these wet days, identification is necessary. I once identified seven different species of flies which hatched in quantity together one afternoon in early May. On this occasion some of the trout were feeding indiscriminately, but others were being very selective.

Though wet days are good for the appearances of duns, the spinners prefer to remain where they can keep dry beneath the leaves of various vegetation. I think, in shedding the dun covering, the fly loses its protection against wind and rain, so they wait until a warm and dry atmosphere gives them encouragement to leave their shelters, and complete their life cycle.

Most creatures have a place of refuge—a home if you prefer it—and this is often borne out forcibly by the action of a trout after it has been hooked. Generally speaking, the brown trout of a chalk stream put up a poor fight for their liberty when hooked with an up-stream cast of dry-fly or nymph, but there are exceptions. Once I was asked if I could explain the fact that trout on the shallows play so much more strongly than those hooked in the deeper water, and why occasionally fish will make a long run either up-stream or down with unusual vigour. If one considers fishing from the angle of the sport given by the fish after it is hooked, then much sport was lost on chalk streams when the up-stream dry-fly and nymph took precedence over the down-stream method of wet-fly fishing, for fish invariably fight better when hooked from a point up-stream of them and the reason is not far to seek.

All trout have a place of refuge, a sanctuary where they habitually rest and to which they instinctively rush when frightened. When trout expect surface borne food the majority move to a feeding place where they can lie in a current and intercept the insects as they drift to them. At times these positions may be ten or twenty yards or more up-stream of their sanctuary. When we find trout rising merrily we move to a place from which it is possible to present the artificial in the easiest manner. This may be anything from ten to twenty paces down-stream. Perhaps the very point we have chosen to stand to make our cast is on the bank immediately above the home of the trout we hope to catch. The artificial deceives the fish and he is hooked. Perhaps he is hurt, perhaps only alarmed, but in either case he knows something unusual and unpleasant has happened, and his natural impulse is to go home. Down-stream he comes with a rush to his holt just beneath us, and there he splashes, rolls and bores, tethered and puzzled, until the landing net slips beneath him.

A poor fight, you say. Yes indeed it was, but how different it might have been had that fish been hooked from a point twenty paces up-stream of his rising position, or had his home been up-stream instead of down. This explains why fish fight so much better in the shallow water, for often the sizable trout are feeding in positions that are a considerable distance up-stream, or down, of their sanctuary, and they make every endeavour to get to it.

So when someone speaks of the poor fight put up by this or that fish, caught with a dry-fly, or when they look scornful when you tell them you killed a 3 lb. trout in less than a minute with a four-ounce rod, please do not think the chalk stream trout have no spirit, strength or desire to fight. It is just that when you net them you are standing close to their home and they have no wish to go elsewhere.

Nymphs and nymphing trout can be put into two categories. I believe that the active nymph is in the stage in which it is most likely to be taken by trout, and the active nymph is within a day or less of hatching into a dun. It is the perfectly mature insect in a stage when it is anxiously waiting for conditions which will enable it to complete its life cycle.

Though I call them active nymphs, perhaps restive

nymphs would be a more fitting term to use. Their underwater life is over, no food is required or taken, they are just
waiting until they feel that weather and water conditions
will be satisfactory for them to mate, and for the females to
return to the river to lay their eggs. And so many of the
mature nymphs come out from their shelters where they
have lived in security from the fish, to move about the river.
Some climb up the vegetation to cling and crawl about on
the fronds just beneath the water. Others delight in swimming from one weed-bed to another, and still more move
freely about in mid-water, swimming towards the surface
and returning to the gravel of the bottom once more. Most of
them will wait until the cool of approaching evening, or
possibly until after it is dark, before they decide it is time
they left their liquid home for ever.

Throughout the day this restless movement of nymphs
will continue and who should know more of these habits
than the trout. Indeed they do know of them, and so station
themselves in places where, without much trouble or
exertion, the nymphs can be intercepted and eaten. Often
enough, the best position for a trout to see and to catch these
active nymphs is near the river bed, for there he can obtain
an unrestricted view of his surroundings. Unless something
very attractive passes above him on the surface he is quite
content to take only the insects which move in the small
area around him, an area which often enough does not
exceed an 18 in. radius. There he may lie throughout
the day and unless some big fly, fluttering and gyrating,
passes above him on the water he will not rise to the surface
to give an indication that he is feeding.

It is useless to try to tempt such a fish with an artificial
nymph fished just beneath the surface, or to cast a dry-fly
over him. He us expecting his prey to be moving near to the
river bed and if he is to be deceived and caught the presentation of the artificial must be made in a natural manner.
It is necessary to get the nymph down and into his radius
of vision and so the artificial must be constructed accordingly. The trout's view of a swimming nymph is to see the

insect with its legs tucked in close to its body, and with these active nymphs the wing cases are a very conspicuous part. So in tying a representation these two features should be considered. Hackles to suggest legs are unnecessary but the body should be built so that the thorax is accentuated. It should be remembered that the nymph has to sink rapidly through the water. I find it pays to dispense entirely with tying silk and to use, instead, a very fine copper wire of a colour which merges with the general tone of the tying. With this fine wire I build up the body and thorax to make a base for the dressing and to give additional weight to the hook. In fishing the representation, the finer the cast, within reason, the better.

Much has already been written of the hatching nymph and of the method adopted in the presentation of an artificial which imitates it. I feel sure far more is known of this type of fishing than of that of the active nymph, for the late Mr. G. E. M. Skues was an artist in catching fish with representations of hatching nymphs, and in his book, "Nymph Fishing for Chalk Stream Trout," he has passed on much of the knowledge he gained.

Whereas the active nymph is in a stage of development about a day or less before it hatches to a dun, the hatching nymph is the creature in the very last few minutes of its underwater life. Indeed, the insect has risen to the surface of the water and is making final preparation to start its short life in the air. The time has come for it to hatch: between nymphal shuck and the fly there is a division of air, the skin is loose, and along the sides of the creature the breathing appendages and swimming aids have ceased to function. Even if the insects make a mistake in thinking that conditions are satisfactory for hatching, and on arrival at the surface find the film of the water to be of too great a strength and density to be easily broken, they have in a manner of speaking already burnt their boats. Though they may wish to return to the river bed they cannot. They have become buoyant with the air, or gas, that is secreted between the nymphal and dun skins and in the wing cases, their respiratory and swimming aids are useless, and they are forced to remain at the surface to become a plaything of the vagaries of the current, and be at the mercy of a rising fish.

Mr. Skues wrote of these hatching nymphs as being inert. I cannot agree with this, for, to be inert, would mean that they show no sign of movement. Even though these hatching nymphs will float for many yards in the surface film, they make repeated efforts to hatch and, in doing this, there is considerable movement of the legs and setæ as the insect does its utmost to thrust its thorax through to the air. At these times trout are expecting flies to hatch, and many have positioned themselves at points where both nymphs and flies can be intercepted. This positioning is done with the knowledge that the food is to be obtained near, or on, the surface of the water. The trout poise themselves at a point a few inches beneath the river level and there wait expectantly.

But now the eyes of the fish are focused in an upward direction, the nymph is no longer active and likely to come speeding up from below or to one side or the other. There they lie with tail and other fins moving in a leisurely way. They know the insects are helpless and just wait until their prey is borne to them by the current. Occasionally their heads will break the surface as they take nymphs struggling to hatch, but usually a hump or bulge is all the indication the angler can expect to see, as nymph after nymph is taken.

From all this it is obvious that a heavily-tied artificial could not be fished in a manner expected by the trout, so a representation must accordingly be made. The artificial must be cast so that it floats just in, or beneath, the surface film, and no drag or movement should be given to suggest life. Indeed such movement is far more likely to scare than to attract, as it is unnatural. Instead the artificial has to be constructed so that it appears to be translucent and gives the effect of air beneath the nymphal shuck, and with a few hackle fibres which can play in the water to suggest movement of legs and setæ.

116

The manner of casting must also be changed, for instead of pitching the nymph into the water so that it sinks rapidly, it should be cast almost as delicately as a dry-fly. Then as it drifts gently with the current over the trout, his eyes are attracted by the moving hackle fibres. He sees a certain translucency over what he is deceived into thinking is a living insect. He senses it is different from the scores of empty shucks that have been passing over him, and often, without the slightest suspicion, he moves forward and upwards to take.

Personally I think there is far greater skill required to deceive and catch a trout with an artificial nymph fished up-stream than there is with a dry-fly, though I know this is contrary to many of the views expressed throughout the years. One thing is quite certain, successful nymph fishing depends to a very great extent on good eyesight and in this alone it must take some precedence over the dry-fly method. In addition, a far greater knowledge of both trout and insects is needed, for one has to be able to see the fish, know what he expects to see and, what is still more important, to have some idea what the trout will do when he does see his prey, or the artificial.

In the clear waters of the south country streams, trout have become very shy by the end of June and as the season advances they get even more wary. Many feed exclusively on sub-aquatic food, and only by nymph fishing can one expect to get sport. Though in my experience very few anglers fish with a nymph, it is, nevertheless, a most fascinating method of catching fish, but I feel so little is known of the underwater activities of trout and insects that many fishermen are not aware of the chances that are presented to have a day of enjoyment.

With nymph fishing the great thing is to be able to see the fish before it has a chance to see you, and only by a stealthy approach along the river bank and a very close scrutiny of the water can this be accomplished. But when a fish has been located, it pays to study it for a little while before attempting to make a cast, for often enough a few moments of intent observation will provide a clue, and possibly a solution, as to which type of artificial is likely to be the most successful.

Sometimes the short study will prove that it would be futile even to attempt to present an artificial, for, though a trout may be lying out in full view to you from the bank, this does not always mean that he is on the feed. Neither does it prove that the fish is not aware of your presence. A good plan is to take note of the position of the trout in the water: at what depth he is lying and if he is in a place where he is likely to see nymphs; and by the nature of the water, fast current, slow, medium or sluggish, to determine the kind of nymph most suited to the environment. Study his tail and his pectoral fins; his movements, if any, from side to side, forwards or backwards; and the eagerness with which he takes his prey. If both tail and pectoral fins are moving evenly and rapidly, then he is on the feed. Even though he makes no appreciable move in any direction his eyes are focused for movement of insects. If the tail only is moving and the fins appear to be held rigid, then the chances are that he has discovered your presence and is alarmed. A discreet withdrawal for a few moments is to be recommended. Should neither tail nor fins be moving, then move on and look for another, for this fish will not be interested in anything you are likely to offer, he is resting—in a state of coma—and almost oblivious to what goes on around him. Depth of water can also be a clue as to the kind of nymph a trout is expecting to see. Some days you may find fish feeding near the surface, on others near the bottom. Occasionally, no two fish are feeding in an identical manner, but it all has to do with the movement of the insect life.

Watching a fisherman having difficulty with a weeded trout brought to my mind an incident about which I told him after he had with much patience persuaded the fish to leave his fancied security. It was on one of the upper reaches during an afternoon towards the end of May. Mayflies were hatching well, fish were feeding and the weather was warm and pleasant. Fifty yards up-stream of where I stood on the bridge a fisherman was preparing to make a cast over a trout which was rising on the far side of a bed of crowsfoot. Only the day before I had watched in that very position and knew it to be a fish of about one and a half pounds. I saw the rings spread as it rose to another fly. It rose again and I had every reason to assume it was the same fish and one likely to be caught.

The angler made his cast and I thought as he did it that he would have stood a much better chance of hooking and landing the fish had he fished for it from the opposite bank. However, I had no time to make this suggestion. The artificial—a Mayfly—dropped perfectly a foot above the position of the fish. It drifted down-stream and was instantly taken. I saw the line tighten as the fisherman lifted his rod, and then I saw the cast draw under the bed of weeds. The rod bent as pressure was applied to get the fish moving, but it was too late. The trout had taken control and was well and truly weeded. I went up-stream to help if it was possible but, though between us we tried every known method of extricating a fish from weeds, it was no good. The trout—though at times we could feel it struggling feebly—remained fast. For nearly ten minutes we tried—slack line down-stream, slack line up-stream, handlining and rod pressure—and then suddenly it occurred to me that the fish might possibly move if pressure was exerted from the opposite bank of the river. I explained the idea to the fisherman who was very anxious to get the fish if he could, as I told him of the size I thought it to be.

At this point the river is only a matter of about forty feet wide but it was much too deep to wade. So I suggested that we removed all the casting line from the reel, snap it at the backing and then tie a stone on to the freed end and throw it over to the other bank. This was agreed, and I found a stone weighing about half a pound and tied it on to the line. It was decided that I should go to the bridge and cross to the other side—he was to throw the stone across and then follow me with the rod. Before leaving I told him to throw the stone well into the air so that it would pick the line up off the water and weeds on his side. I ran down-stream to the bridge, crossed it and had just got opposite to him when I saw his arm go back as he prepared to throw over the stone.

Whether he thought the casting line was a hundred yards or more in length, when in fact it was little more than twenty yards, or whether he thought the river was wider, I do not know. I expected him to cast the stone so that it just reached the other bank. But up it went, like a rocket, and came whizzing over the river like a cricket ball thrown from the deep field. Sixty feet in the air it went with the line trailing behind like a long tail. Then suddenly the line tightened, and with a jerk which must have nearly dislocated its neck, a five-inch yearling trout hurtled out of the weed-bed and sailed away on the end of the cast and out into the meadow beyond. Which was the most surprised—the fish, the fisherman or I—it is difficult to tell, but never before, or since, have I seen a hooked trout come out of weed more quickly.

When limited to a single bank of a fishery there is nothing one can do but to use one's skill and hope for the best, but where there is choice of either side then it often pays to give a little thought to the subject before starting to fish up a reach with a dry-fly. Nine out of ten fishermen are right-handed—that is, they hold the rod in their right hand while casting a fly. Occasionally there is one who is equally good with either hand, and then there are the few that are naturally left-handed. If there is a choice of banks, the right-handers invariably choose the right one, as they find it to be easier to cast a fly under their own bank and the left-handers usually fish the left bank for the same reason. The ambidextrous fisherman is greatly influenced by weather conditions. But in nearly every case I see, if there is a wind blowing up-stream, or dead across the river, the angler chooses the bank from which his casting can be helped by the wind—that is, with the wind blowing from behind him.

Granted, a following wind is a great help in getting out a long straight line, and is less tiresome than having continually to drive line and fly under a breeze, but in dry-fly fishing the idea is to present the fly in a manner that appears natural to the fish, and not to accomplish any long-distance casting accuracy. The mere placing of an artificial fly on the surface in front of a rising fish is not enough, the fly must float towards the fish in a natural way. If a fly has to float any appreciable distance with a wind blowing from angler to fish, this natural presentation is impossible.

Trout are easily scared, and even if the surface of the river is broken and disturbed by the wind, a dragging fly will still put a fish down. An angler would be horrified if he allowed his fly to drag over a rising trout where the water was calm and unruffled. He would know that such action would scare his fish. Yet the very same thing happens time after time when that same angler is casting with a following wind, but the drag remains undiscovered, as it is concealed by the turbulent surface of the water.

The natural tendency of all natural-floating flies is for them to drift with the wind. If the wind is blowing from your bank towards the other side the majority of the insects move with it across the river. But it is quite obvious that an artificial cannot do this. It is tethered to line and to rod, the result is distortion. The action of the wind tends to straighten cast and line, the artificial, instead of drifting naturally on the surface, is checked, and in consequence is dragging against the action of wind and water. The fish know at once that something is wrong.

There are times when there is a strong wind up-stream, when flies, instead of floating with the current down-stream, are being swept with the wind in an opposite direction. Here again, an unnatural presentation of an artificial can be made if a fly is cast up-stream with the wind. The action of the current has a strong pull on the line and cast, and the fly just sails along against the wind like a yacht on tack. But much of this drag effect can be overcome by casting a very slack line and cast, as one would do when placing a fly on the opposite side of a fast current, but few anglers do it. If I have a choice of banks I much prefer to cast against the wind, for when once the fly has been presented to the fish it floats in a natural manner and is likely to deceive. I demonstrated this successfully one afternoon, when one of the members and I were after grayling. We were fishing opposite to each other, and both were using the same pattern of fly and 4x casts. He fished with the wind, and I dead against it. We were at it for about an hour, first he covered the fish, then if he failed to rise it, I made my cast. He got four grayling and I seventeen.

So many people think that the art of dry-fly fishing is a fairly modern accomplishment that I would like to know for certain just which came first, the wet fly or the dry. This is how I think it may have begun. The first fly fisherman would have been a student of Nature—he would be well aware of the fact that trout took various flies from the surface of the water. To me it is obvious that his first thoughts would be to use a natural fly—a big fly such as the Mayfly, Sedge, Alder or Stone Fly—and impale it on a bare hook and that the method was to allow the bait to float on the surface of the stream in a manner which represented the natural insect. In fact, it was fishing with a dry fly.

But it was found that these natural flies soon became battered and broken when being presented, or when a fish was hooked, and so the art of making artificials to represent the natural flies was adopted. As it appears that little was known of the underwater habits of our trout flies, is it not probable that the first artificials were made with the intention that they floated on the surface of the water in a crude imitation of Nature? Owing, possibly, to the heavy hooks, casts, lines and tackle generally, the artificial, more often than not, sank beneath the surface and there was acted upon by the vagaries of the currents so that sufficient lifelike movement was imparted to delude a fish into taking it. I would suggest that in this way the art of catching fish with an artificial fly, beneath the surface, was discovered.

All well and good. It was wet-fly—a wet artificial fly. With the tackle to hand, no doubt the presentation of these wet-flies was found to be much easier than the dry. Dry-fly was abandoned in favour of the method which produced most fish. Soon patterns were tied so that they quickly entered the water—tied with softer materials, and arranged in such a manner that wings and legs collapsed and draped themselves round the body and hook. But because fish took these creations, and because originally they had been made to represent flies, it is perhaps a fallacy that they were, and still are, taken by fish in mistake for the creatures of which they are a copy.

As time went on, wet-fly patterns were constructed (some are still in use to-day) to represent nearly every insect one is likely to find on, or near, a river. At some time or another they have all caught fish; yet, how many, I wonder, were taken because the fish were deceived into thinking they were flies. Some well-known authors still persist with theories that natural flies become storm-tossed, drowned and waterlogged and are then carried floating down-stream in midwater. I wonder why? Though many millions of flies may hatch from a river each year, it is but few that are likely to be seen beneath the surface in winged form by trout, and these only because they choose to return to a place below the water to complete their life cycle by laying eggs.

Most of the trout flies spend nearly all of their existence beneath the water, in larval or nymphal form, yet, when once they have reached the open air and are transformed into winged flies, it is very, very difficult to make them sink. No storm or broken water is likely to submerge them indefinitely and trout have no opportunity to form a habit of taking them in a drowned or bedraggled state. If the old people were deluded into thinking fish took the wet-flies in mistake for the creatures they were supposedly representing, it is time, with our advanced knowledge of natural history, that the modern generation knew differently. I think trout are deceived into taking a wet-fly pattern not for what it is made to imitate, but in mistake for one of the many sub-aquatic creatures they are used to seeing, and into which shape, coloration and the action of stream and rod tip has transformed it.

10   CONVERSATION PIECES

Whenever a new trout season starts I know I shall be asked many questions concerning the insect life of a chalk stream. Some of the questions are, to my mind, humorous, and often require diplomatic answers. The names given to artificial representation of many river flies are fancy ones, and it is nothing unusual for people to speak of hatches of this, or that, fancy fly. Most of the natural insects, at least, the most important ones from a fly fisherman's point of view, are grouped under the heading of ephemeroptera. They are flies which have a very short existence in the air. In their varying stages they are referred to by fishermen as larvæ, nymphs, duns and spinners.

The name spinner, I must admit, is confusing because the name is also given to many other lures. Often they are made of bright metal. I was talking to a lady one day. She was interested in fly fishing, and we were discussing the merits of the various rivers in the locality, and their value as fly-fishing waters.

"Tell me," she said, "this Upper Avon is dry-fly water, isn't it? But spinners are allowed, aren't they? I ask because last evening I heard a fisherman telling his friend that he had killed a brace of good trout on a rusty spinner. But why

124

a rusty spinner, Sawyer—have they more attraction than the silvery ones?"

Well, in this case I thought the least said the better, and gently explained that a Rusty Spinner is just a fisherman's name for an artificial fly. I quickly drew her attention to a pair of coots, on the far side of the river. Another question followed.

"Coots, did you say? Are they the Black Drakes which dance over the river in the Mayfly season?"

Those are but two examples of questions I have to answer, but there are many others of a scientific nature. Is this an Olive or an Iron Blue, a Pale Watery or a Spurwing, a Blue-winged Olive or a Turkey Brown? Which of the Mayflies is this? Is it a male or a female?—and so on. I know these questioners cannot be expected to do the things that I have done to find out the identity of river creatures and in getting my knowledge I know I must often have given people very good reasons to think I was a candidate for a lunatic asylum.

One afternoon I was studying some nymphs. I was watching them as they moved about the river bed, on the vegetation, and noting their progress as they came to the surface to cast their nymphal shucks, and hatch. I was very intent on my subject and was lying at full length on the river bank, with my eyes within a few inches of the water. Time passes very quickly at this occupation. I expect I had been there quite an hour when in the bushes behind I heard whispers, and caught part of a sentence, ". . . waiting for the water to draw him into it. What shall we do?" Quickly I looked round. There stood two airmen, both with horrified expressions on their faces, which changed the moment I grinned.

"Hello," I said, "I didn't hear you come along. Did you wonder what I was doing?"

"Wonder," one of them replied, "I should think we did —both of us thought you were going to drown yourself, and we were going to grab you. What were you doing?" I then explained, and indeed showed them the subject in

125

which I had been so engrossed. Then they, too, became interested and I had many questions to answer.

One gets disappointments, too. Another day I was studying a rare nymph and was anxious to see the fly which hatched from it. I could see the insect clearly as I lay on the river bank, and knew it was going to hatch. But a considerable time passed before it came to the surface, and then it took me by surprise and hatched immediately. Grabbing my cap, I scrambled up and made a swipe with the cap to catch it just as a swift whistled over my head and beat me to it. A fraction of a second, and I believe I should have got both bird and fly. I have hated that swift ever since, for I have not identified that species yet.

My craze remained with me during the war, and whilst in the War Reserve Police I had some very embarrassing moments in consequence. One evening, while cycling near the river, I spotted a male spinner I wanted to complete a series for my collection. I had found it to be quite easy to catch flies in my policeman's cap, but this evening was wearing a helmet. Anyone who has tried to catch a fly in a bucket will know how difficult it is. Three times I got it in the helmet and stopped with one foot on the bank to take it out. Each time it eluded my fingers and sped along the road. I had to pedal hard to overtake it and then made a sweep. My helmet slipped from my fingers and rolled into the middle of the road just as a staff car full of Army officers came round the corner.

"Hard luck, constable, tighten up your chin-strap," one of them shouted as the car pulled up to allow me to retrieve my headgear. Yes, indeed, hard luck I thought, as my fly fluttered from inside the helmet and sped away across the countryside. I jammed the helmet on my head, saluted and passed on, thinking of the fly that should have been in my matchbox.

But this habit of keeping flies in a matchbox caused me some more embarrassment one evening. I had caught a dozen Sherry Spinners and had put them in my matchbox to study under the microscope when I finished duty. I had forgotten them when later, in the office, the sergeant asked me for a match to light his pipe. As he put his finger into the box one of the flies crawled up it and three more flew past his nose. Of course I explained. My explanation, however, left him with an uneasy feeling that I was not all there, a suspicion indicated quite as much by his pitying look as by what he said. I hastily withdrew.

How surprised a fisherman would be if I said in reply to a question, "Oh! yes sir, that's identical with the fly I got in my helmet in '42." He would probably think I still had some bees in my bonnet.

Someone once asked me if I thought an otter could swim faster than a trout. I replied that I was very doubtful and asked the reason for the question. Apparently my questioner had thought that otters chase the fish all about the river, much as a dog would chase a rabbit in a field, so I told him a little about them. Otters certainly catch trout and other fish very successfully, but it is seldom that they chase them far to do it.

Otters are fond of trout; they are also fond of pike and eels and I think that by killing pike and eels, the otter makes up for what damage he does by taking trout. He does far more harm in a salmon fishery, where he often kills many more fish than will satisfy his hunger. But I speak of a trout stream, and here, no matter where it might be, the otter seldom pursues his prey for any great distance, and if you know the habits of fish, the answer is quite obvious. He has no necessity to do so. You try chasing a trout or a pike yourself and see what they will do. The first impulse of either is to make for cover and the closer this is the quicker he is in it.

Otters catch and eat a large number of eels, and this is quite easy to understand, for eels feed mostly on the river bed where they can be seen and taken with little trouble. But eels are not always procurable by otters through the winter months. They lie dormant, and in many cases are buried deep in the river bed and banks. So the otter turns his attention to trout and pike. Let us imagine an otter diving or swimming into, or through, a pool. All the trout and pike quickly dart to cover—into a weed-bed, under a stone, in a crevice, or under a bank—in most cases (like the proverbial ostrich), they feel quite secure if their heads and eyes are hidden. Their tails, and often three parts of their bodies, are plainly visible to eyes much less keen than those of an otter. He little cares if it is the head or tail that is showing, as long as he can get a grip with his sharp, pointed teeth. The same scurry for shelter would occur if you or I dived into a pool or swam through a reach of water. I have caught many trout and pike with otter bites near the tail end, caused when they managed to wriggle free from the sharp teeth.

Occasionally, roach and dace are taken in the same way, but I seldom find a grayling that has been killed by an otter. The reason is that grayling rarely seek cover, no matter how they are frightened. In this case, if an otter wants to catch one he has to swim until he tires it out.

The way in which trout go to cover even if humans dive into a pool was borne out forcibly by a man I knew a few years ago. He knew the habits of trout as well as any otter, and he could swim almost as well as one. He was swimming in a deep pool near his home and a dozen or so people from a nearby town stopped to watch him. He was an excellent diver and comments passed frequently amongst the on-lookers in praise of his underwater evolutions. They were overheard by the man's friend who was also watching, and he led them on.

"Swim," he said, "he can swim like an otter. Why he can travel fast enough to catch a trout."

But this the townsmen would not believe and several offered to bet that he could not catch a fish for them at that moment. Calling to his friend in the water he told him of the offers to make wagers and both accepted. The swimmer dived deep into the pool and disappeared. In a few moments he surfaced with a struggling trout across his mouth which he brought to the bank and laid in front of his audience. To say they were astonished is to put it mildly and some thought there was a trick in it. Bets were made that he would not repeat the performance, and down he went again. He re-appeared with a second fish in his mouth and the wagers were won.

What he had done is now perhaps common knowledge. I have done it myself. He knew that all the trout had sought cover under a ledge at one end of the pool. He knew also how to take a trout from such places with his hands, and had done so. It was the work of a moment to transfer the fish to his teeth to bring it to the surface. If it is as easy as that for a human, how much easier it must be for an otter.

Many of my conversations beside the river illustrate the hold that fishing has on a man, no matter what his cares and responsibilities. A chance encounter with an Army officer proved a case in point.

His face, though aged and lined with worry, was familiar, and as my thoughts raced back through the years, I remembered his name.

"Captain — —," I started to say, then stopped at the look of amusement he gave me.

"No, no, Sawyer," he hastened to say, "not now. I've been promoted since I last saw you. I am a major-general."

I apologised and then congratulated him. I felt pleased to think that with his rise in fame and fortune he had not forgotten his fishing days of long ago—there he was, just as excited about the prospects of catching a trout as he had been then, when duty and trouble weighed lightly upon him. I did not waste time in asking questions. I thought he wanted to fish and to enjoy his day on the river. Trout were already moving, time enough later on to talk of the intervening years. So, putting him on to a brace of rising fish, I left him.

The general, when greeting me, had said, "You haven't changed much, Sawyer." Well, perhaps not, in face or figure. But as I wandered down-stream I knew I had altered con-siderably in other ways. Even as he had risen in his profession I felt I had risen in mine, for I had learned much in fifteen years. But how different were our two lives. His had been a study of peacetime military organisation, and then of war; mine had been a study of Nature, to find how I could help to create and to foster. I began to wonder which of us was the more content.

There is something about a river which holds a peculiar fascination for many of us, for there, gathered together and part of the living element of water, are thousands of different beings. As individuals they are perhaps insignificant, but when in numbers and in combination with others they provide a beauty and interest not to be found elsewhere. To-day the water meandered peacefully down the lower part of the valley between beds of flowering crowsfoot, while on either side the meadows golden with buttercups and marigolds swept back to where the leafy trees shaded the slopes and the hillsides. All around was the music of birds, the deep mellow tones of the blackbirds mingling with the songs of the warblers. Bees buzzed around the blossoms, and river flies danced in the rays of the sun. Occasionally the surface of the river was broken and as the rings made by a rising fish spread gently towards the banks all else could be forgotten by the fisherman.

All else could be forgotten. Was it to catch fish that the general had come to the river, or was it for another purpose? I came upon him later that day. Without his knowledge, I studied him. There he was, crouched low behind a fringe of willow herb with rod in hand. His eyes fixed intently for a few moments on a spot under the opposite bank, then they would sweep up-stream and down, watching the activities of the swifts and the swallows, then to rest a while on a vole busily munching a strand of weed on a tree-root, on to a wagtail flitting about on a weed-bed to take the hatching duns. And he was listening to the young starlings calling to their parents from amid the blossoms on the hawthorn bushes.

Occasionally he would glance to the artificial he held in his hand, and always his glance would return to that spot under the other bank. Though I could see his actions and his interests, I could not read his thoughts. But as he turned in my direction the expression on his face was enough. No longer was it the face of a commander of men, it was the face of a boy—enraptured. No thoughts of war and destruction could bring that peaceful repose. He was content and happy. I discreetly withdrew.

It was evening when I saw him again, and he was preparing to leave. "Sawyer," he said, "I have enjoyed my day, but how differently to what I expected. I came here this morning to be quiet, so that I could have a chance to think over an important paper I have to write for the powers that be, but to-day this stream and valley have been far removed from war. My report is still unfinished, I have not attempted anything and, for the first time in many months, I feel refreshed. Sawyer, what is it about a river that makes you forget?"

Well, there was no need for him to ask that question. He should have known, even as I did. That day he had been near to Nature, he had enjoyed it subconsciously, for, without knowing why, his eyes and ears had been seeing and hearing some of the things they were meant to see and to hear, and they had communicated a feeling of peace and happiness to his tired brain.

Not every meeting is as simple. One afternoon early in May last year, a fisherman came to my cottage in great excitement.

"I've had a wonderful experience on the top shallows," he greeted me. "In one place there, I have cast flies and nymphs to rising fish for more than two hours. They were still rising when I left them to come down to you—there must be dozens of them, all feeding on nymphs on the top of the weed-beds—great, big fish, too. I think some are grayling, they look silvery. They are in the fast water as it breaks over the weed-beds about half-way up the shallow. Do tell me what you think they are taking. I must have risen a hundred yet I haven't hooked a single one. Never have I seen so many fish rising as madly as they are, and in such a small area."

Well, here I was faced with another question which needed

diplomacy. How could I tell this excited angler that he had been wasting his time fishing for a big shoal of spawning roach, for undoubtedly that is what he had been doing. Early May is the time of year when both roach and dace spawn in south country streams, and only that morning I had seen this big shoal on the shallows he mentioned. Even then I was putting some extra winging on to the front part of a pike trap so that I could use it to catch up some of these spawners and destroy them.

After thinking it over for a moment, I decided to tell him the truth. I did not want to embarrass him, but I was afraid he might tell his story to someone else who was aware of what was happening and then they would both think very poorly of me for evading a truthful answer. So, trying not to cause offence, I explained. I then told him of my intention. He took it very well, and laughed at his own folly.

"By Jove, what a fool I have been!" he chuckled. "No wonder I couldn't hook any of them. Can I come and watch you, I'd like to get my own back on the beggars?"

"Of course you can come," I replied. "It won't take long to finish this trap."

A few minutes saw the job completed, and, carrying the pike trap between us, we set off for the top shallow. As we approached, I could see the shoal was still there and at once it was apparent how easily one could be disillusioned into thinking they were feeding fish. Splashes and waves were churning the water into a turmoil—tails and dorsal fins were breaking the surface, and every now and then the light reflected from the silvery sides of the fish, as they turned and cavorted, and as some hurled themselves clear of the water. They were spawning where the main current of the river was sweeping over a bed of ranunculus in a well-aerated stream. Ten yards down-stream, the run formed into a narrow neck between two big weed-beds. Below this was the deep water—the home of the shoal to which, when disturbed, I knew the roach would return. Wading out stealthily, I set the trap in the narrow neck, and into this, with the two extra wingings I had attached, it fitted snugly.

132

Soon it was secure against the rushing water with its mouth facing toward the shoal.

Shoals of roach and dace are like flocks of sheep. They will all follow a leader. This shoal was no exception. The roach became alarmed as I walked into the shallow up-stream of them, and together they bolted down the run. I had estimated the shoal to be about one hundred, and most of them went into the trap—a few turned once more up-stream and others went into the weed-beds on either side. Dragging the trap to the bank we found on counting the catch that we had sixty-three. All were ripe for spawning.

Amongst the scales on most of them, more especially the males, were the little white tubercles that are characteristic of both roach and dace at spawning time. These tubercles interest me, for it would seem that they exude something which completely destroys the slime on the scales for a period just before and until spawning has been accomplished. When handled, the fish have a feeling of dryness, of rigidity, and obviously it has something to do with spawning. My companion was most interested and astonished that he could grasp and hold the fish so easily while they were alive and wet. But though he questioned me concerning them, I was unable to answer with any degree of certainty. I think it must be that the usual slime of these fish could have a harmful effect when so many of them are gathered in a small area of water, as at spawning time. The slime, if mixed in quantity with the milt, or the eggs, might cause bad fertilisation, as indeed does happen, if slime from the backs of trout is allowed to mingle with the eggs when artificially stripping them into a tray.

11　INTERLOPERS

THE Hampshire Avon is noted for its grayling, but I have often cursed whoever introduced them to a river which, in its upper reaches, is so ideally suitable for producing first-class trout and, what is more, for producing the multitude of insects which play such a big part in dry-fly trout fishing. Some people like fishing for grayling to such an extent that they prefer them to trout, but they are very few indeed. It is very difficult to please everyone, and the fact remains that wherever grayling become established it is almost impossible to get rid of them completely. So in the Upper Avon grayling continue to be appreciated by some, and disliked by the majority.*

Though considerable numbers are netted and destroyed, sufficient remain to show up on the spawning grounds. The sites chosen are somewhat similar to those selected by trout. Clean, well-aerated gravel is preferred, and in many cases they spawn on the redds of trout. Compared to the time taken by trout, the fortnight which completes the grayling's spawning season is very short, and this is usually during the latter part of March and early April. The actual shedding of eggs by an individual is done in about two days, and it would seem that all fish have the urge to spawn

* See Introduction, p. xiv.

within the two-week season, when river and weather conditions are most favourable.

Though occasional isolated pairs may be seen, grayling prefer to spawn in company. There is no true pairing, and a female may be attended by as many as a dozen males. This last statement may account for the fact that many writers say grayling change colour at spawning time, taking on a much darker shade, and when viewed in the water appear to be almost black. It is perfectly true that both male and female change from their former hue, but only the males change to the darker shade. The female takes on a colouring that is much lighter than her former pigmentation—a shade which merges so well with the clean gravel of the river bed that the keenest observer can easily miss her presence. This different colouring of the sexes makes it a simple matter to tell the two apart when on the spawning beds. The dark coloured males are quite easy to see, as they dash madly hither and thither about the river, chasing and fighting each other, and following the females as they scatter their eggs over a large area, as in fact they do. For, though grayling choose sites that are similar, they make no nest or redd, like trout. Flexions of the body are made, and eggs are strewn all about the river bed, and there they are fertilised by the males as they follow the female around.

The eggs when laid are about one-third the size of those of a trout, and it is estimated that a $1\frac{1}{2}$ lb. fish can shed between 4000 and 5000. Luckily, from my point of view, there is a heavy mortality. Many of the eggs are not fertilised, and others succumb before hatching. Grayling eggs are destroyed in thousands by many different enemies, but even so a considerable number hatch and thrive, to become contestants for the insect life which should feed the trout.

Probably because familiarity breeds contempt, I have little regard for grayling as a sporting fish. I think of them as interlopers in a trout stream and I am doubtful if there is a period during daylight, between May and November, when they will not feed if the opportunity is presented. They are greedy fish and the water of the Upper Avon suits them well. The alternate formation of the deep and shallow river bed is to their liking, and the slow-paced current, passing over the gravel and weed-beds, breeds the food they find most nourishing. Here they feed principally on fly larvæ and shrimps, and so quick is their eyesight and feeding movements that any such creature moving within their range of vision courts disaster.

This water is also ideal for the regeneration of grayling, and each year I have to find means of reducing the numbers which otherwise would give trout little chance to take floating food. Sometimes I net them, at others I use traps, but often where netting or trapping is impracticable, or not to be desired, I take a fly rod, and with a lure of my own construction I fish to kill. Grayling have a colouring which merges well with their surroundings, but it is not difficult to locate a shoal if you know where to look for them. I have found, however, that their eyes are no less keen in looking towards the banks than they are for the movement of insects and it pays to make a very cautious approach. They are less likely to be alarmed if an advance is made towards the sun and where access along either bank can be made, this point is well worth keeping in mind. Once a shoal starts to mill around, it is clear proof that they are frightened, and it might be half an hour, or more, before they settle down once more to feed without suspicion.

Considerable numbers of grayling can be caught with a rod if the right tactics are adopted. Once when giving a demonstration to a neighbouring keeper, I caught sixty-three in two and a half hours, and many of these fish weighed between one pound and two pounds. I could easily have killed a hundred that afternoon, but I got sick of the monotony.

As I have said, I use a lure of my own invention. Knowing that grayling are fond of shrimps, I studied the habits of these creatures and so, when I tied an artificial to represent a shrimp, I knew how grayling expected to see it in the river. I watched shrimps as they went about their everyday life and saw how eagerly grayling took them if they swam into mid-water. Shrimps swim backwards, that is, they use the flanges on their tails in a series of quick flips to propel them through the water. In this way they will glide upwards tail first from a bunch of weeds or from the river bed, and often make a half circle, before coming to rest again. In deep water they may rise two feet or more, and in doing so, form an attraction few grayling can resist.

The lure I use is very simple to make. I use a No. 2 or 3* hook and around this spin a length of silver-coloured fuse-wire to give added weight, and a pronounced humped effect, to the hook, finishing near the bend of the hook, where I leave the fuse-wire dangling. Then starting at the eye I cover all the wire with an even winding of ordinary darning wool, and with a couple of half hitches of the fuse-wire the wool can be secured and the spare ends of wire, and wool, cut away. But the most important part is in selecting the right coloured wool. The best I have found is a wool with a fawn background that has a definite pink tinge. This, when wet, turns to a colour that is very much like a shrimp. I have found that there is no need of anything to suggest legs.

Where trout and grayling are together it is often difficult to tell the difference between their rise forms. I am often asked if I can distinguish between the two but though there are times when I can be reasonably sure which fish is rising I am not infallible, neither do I think there is a man living who could truthfully say that he has not, at some time, been deceived.

No hard and fast rule can be laid down. The rise action of a grayling is exactly the same throughout the season. It will take a Black Midge or a Cænis spinner with the same movement it uses to take a Sedge or a Mayfly, whereas the rise form of a trout will vary considerably with the insects he is taking. Unless one is in such position that one can obtain a view of the fish's head or dorsal fin then it is impossible to be quite sure which fish is breaking the surface.

Grayling rise direct from the river bed, no matter what depth the water might be. The deeper the water the more

*No. 2 is now No. 13, and No. 3 is No. 12

the speed they have gathered before they arrive at the surface and so their heads break through with sufficient force to cause a violent distortion. In all cases the rise form is a kind of hillock surrounded by broken rings. And coupled with this speed of approach is the fact that grayling snap at a floating insect and often several bubbles are made by the action of the jaws as they open and close. Compared to the take of a trout, the rise is a much more rapid procedure and it is quite easy to understand why grayling so often miss a natural insect or, when rising to an artificial, how they will strike the cast with their noses and so throw the fly clear of their mouths. On the other hand it is the general rule for trout to take surface food in a very leisurely manner, for in deep water they usually poise themselves at a point that is seldom more than a foot beneath the surface. When rising they move upwards with a steady glide—just a tilt of the head and a single thrust from tail and fins, brings them to the top of the water. Unless the insect that has attracted them is a very lively one—a creature making some endeavour to take flight or to move across the surface—there is no sudden hit-or-miss attempt, but just a gentle opening of the jaws into which the insect passes. The opening of a trout's jaws just beneath the surface causes a vacuum into which the water pours, and, with the water, any insect it is bearing. So the general rise form of a trout takes on a different shape. Instead of a hillock in the centre with broken rings around it, as made by the grayling as it forcefully snaps the fly on the surface, the form made by the trout has a dimple—a kind of hollow—in the centre, into which the water and fly have disappeared. Then as the head of the fish breaks through to the air, the displaced water spreads evenly away in gradually widening circles.

The position occupied by the rising fish can sometimes be a clue to their identity. Trout will often rise in places for which grayling have no liking. Grayling prefer a clear view all around them, and so they are seldom found rising near, or under, an overhanging bank, in small pockets in weed-beds, or even from the up-stream portion of any river vegetation. They much prefer the clean gravel runs in mid-stream, they are not individualists like trout and several fish will usually be found rising in a very small area. Unless one can actually see the fish, I find it to be very difficult indeed to distinguish between grayling and small trout when both are rising in shallow waters. Both have the same quickness in approaching the surface and in taking the surface food.

Predatory feeding by grayling is a subject which has not had much attention. A letter in the *Salmon and Trout Magazine* of May 1946 quoted an extract from the *Zoologicheskii Zhurnal* to the effect that "13 per cent. of grayling in a sample from Pechora, Russia, had bullheads or minnows in their stomachs (August/September fish). Ten per cent. of grayling had one or more small mammals, almost entirely common shrews (*Sorex araneus*) which they caught at night whilst the shrews were swimming in the river. They were found especially in the larger fish. Water shrews, though present in the river, were not found in the stomachs of the fish."

This statement about shrews is curious as it brings out the fact that the common shrews must have been taken from the surface and not under the water as, though it is a common sight to see water shrews (*Neomys fodiens*) swimming under water I have never yet seen the common one dive from choice. Common shrews are plentiful in this district, though it is an unusual sight to see them swimming in the river, either by day or by night. Another fact is that though there are numbers of grayling in the Upper Avon that are between two and two and a half pounds, it is seldom that they rise. They live almost entirely on food taken on, or very near the river bed. These large grayling—like large trout—need a lot of food, and they may also turn cannibal, but so far I have not had experience of grayling eating mammals.

That they eat bullheads is certainly one thing in their favour, but so far I have only found bullheads and loaches in the large grayling, whereas it is the small fish, up to one and a quarter pounds, which are the chief offenders as far as the trout are concerned. I think it is likely that they only take this sort of food during the latter stage of life when they are going blind, the natural end of all river fish. Slow starvation faces them and, like the trout, they turn to anything that will make an easy meal.

For many years I have studied the habits of grayling, but it was not until recent years that I had conclusive proof that they were predatory. My first suspicions were aroused about fifteen years ago. My brother and I were spinning for pike in a deepish reach of water where I knew there were a couple of small jack. We were sharing a rod and using a five-inch Wagtail. A fish touched the bait, my brother struck, but nothing happened. With the next cast the same thing occurred, and again a third and fourth time. In desperation he gave me the rod. I cast to the same spot and felt something like the take of a small jack, and struck. Nothing happened, and I cast again. Once more came the tug and I struck quickly, and with vigour. This time I connected with what I thought was a heavy pike. It played deep down and gave several long runs and, after a period of perhaps five minutes, I was surprised to see I had hooked a large grayling in the dorsal fin. It was an old cock fish and weighed nearly two pounds.

Two years later, when spinning a dead bait—a four-inch trout—in another part of the fishery, I hooked and landed another large male grayling of similar size to the other one. Strangely enough this was also foulhooked, but in the back of its head. As I had had no previous experience of grayling taking small fish, I came to the conclusion that these two fish had tried to take one of the swivels on the trace, mistaking it in the dirty water for a shrimp being chased by a small fish. I am not quite so sure now that I was right in assuming this, since I have had proof that large grayling, especially males, will eat small fish during the winter months. Both the above incidents occurred during February.

One day in January 1942, the river being exceptionally low and clear, I fished a few pools where I could see grayling. I was using a lure of my own invention which represents nothing in particular but might be taken by grayling for a shrimp or a hatching sedge, more probably the former as sedges seldom hatch in January on this water, a fact which is no doubt well known to grayling. I caught several good fish, including one of 1½ lb., an old male in bad condition and partially blind.

As is my usual custom, I made autopsies, and in the old male, amongst a collection of shrimps, caddis, snails and worms, I found a 2-in. bullhead which had apparently only just been taken. I recollect mentioning the matter in a letter to Mr. G. E. M. Skues, who thought it very unusual, but since that date I have examined other large grayling during the months of January and February, which contained both bullheads and loaches.

Grayling have always been a subject for discussion, and lately more and more people are beginning to realise that where these fish have increased the trout fishing has suffered. My experience on the Upper Avon is the same. Though in many reaches there are plenty of good trout, it is seldom that they rise, yet scarcely an hour of daylight passes during the trout season without the surface of the water being broken by the rise of a grayling. Most of these are small fish, hatched in early May of the previous year and are about 8 in. long.

The food of grayling is similar to that of trout, and a hatch of fly is eagerly taken. Where the hordes of grayling are in evidence few fly are allowed to leave the surface of the water. Many are taken in nymph form as they rise to the surface to hatch, and others are taken in their late active nymph stage. Small fly which hatch from nymphs of the swimming class, though plentiful in their early larval existence, are scarce in their more active, latter larval and nymphal stages. Though a good hatch of fly may occasionally appear, the trout are greatly handicapped. Grayling are much quicker feeders and possibly have keener eyesight, with the result that flies are taken from in front of the trout's nose before it even realises they are there. This state of affairs is most unwelcome to a trout and, though it may try for a little while, it finally leaves the field to its more active competitors and retires to a place where, even if the fly is less abundant, it can feed in peace.

One year I proved beyond doubt that the continual activity of the small grayling will have a bad effect on rising trout and that those which would rise are so harassed that they turn to bottom feeding. I netted a pool at the tail of a long, shallow reach. Grayling were there in great numbers and I had estimated there were at least two hundred of all sizes. I had also seen a few trout. My estimate was a little exaggerated, as I think we caught all there were in the pool and they totalled one hundred and sixty odd. We also caught seven trout. All the grayling were killed, together with three of the trout, as I wished to make examinations of their stomachs. The trout were of different sizes, one a two-year-old of $\frac{1}{2}$ lb.; one about 1 lb.; and the other $1\frac{1}{2}$ lb. All were in good condition. I selected a dozen or so grayling of all sizes up to 2 lb. On all all of them I made autopsies.

The results bore out what I had long suspected. The smaller-sized graylings' crops were crammed full of fly life, from larvæ to duns and spinners, together with a few caddis and shrimps, these latter being in the majority in the larger fish. The trout were also full, but with food of an entirely different character, the great majority of which was animal life found in or near the bed of the river. Shrimps and snails were most in evidence, together with fly larvæ in the very early stages; the large one had a minnow and a small cray-fish in addition. Though I cannot be positive, none of the trout as far as I could see, had a fly of any species.

These observations bore out my suspicions that the trout had become bottom feeders but when, two weeks later, I was on the spot during a good hatch of fly, fish started to rise as they had always done, and now it was trout which were in the majority. I counted over twenty of all sizes, including one of 2 lb. which was killed the same after-noon by one of the fishery members. The trout rose well and with confidence, and I was surprised to see such a number in the pool. As a result, it was decided to make a wholesale slaughter of grayling throughout the fishery. Weather and water conditions were ideal and each reach was thoroughly netted. Many thousands of grayling were killed.

The following season proved to be a record for the club, when 1875 trout averaging over 1 lb. were reported killed in the six miles of water that made up the fishery. For the first time the tally of trout far outnumbered that of grayling. Fly of all species were plentiful and the trout rose well. That grayling give good sport throughout the trout season and sometimes far into the winter months, no one who has fished for them will deny, but if a preference is to be given to trout then the grayling must be killed, or at least checked, so that they cannot outnumber them.

Each year as spring approaches the minnows put on their new dresses and come sallying forth from the places in which they have spent the winter. Once again they need food and with the rising temperature of the water the creatures on which they feed will also be awakening. Gone will be the drab colouring of winter which camouflaged them so successfully in the mud of the ditches and spring holes, or which blended so well beneath the big stones and other debris. Their hibernation is over and ahead is the excitement of spawning.

Minnows seldom feed during the winter months, and seldom do they move far from the places they select in which to lie dormant during the cold weather. As boys we used to call them soldiers, this probably on account of the brilliant colouring of the male, and of their habit of moving along in great armies at spawning time—a time, I suspect, when we were most familiar with them. Indeed, at spawning time, soldier is quite an apt name, for the bluey-green back, with crimson, gold and carmine showing on sides and stomach, together with the white lacy appearance at the base of the fins, has some resemblance to the splendour of the dress uniforms of years ago.

I like to watch minnows in March and early April. To me they are very interesting when they congregate. It appears like a rally or an assembly of the tribe, for an initial party may start with a few dozen yet, at the end of an hour or two may swell to ten thousand, or more. Most of my observations of minnows have been carried out in spring-fed ditches and streams. In some, where the bed has been muddy, the water has been perfectly clear at the start of a gathering, yet soon, with the movements of little bodies bursting out from their sleeping quarters and with the tribe as they move up-stream and down, the water becomes so dirty that further observation is impossible. It would seem possible that these little fish intentionally cloud the water to screen their movements from enemies which may be watching from above.

But I have watched other gatherings where the stream-bed has been gravel and where here and there big stones and other debris has been strewn about. Here the assemblies have been of greater interest, this, possibly, because I have had a better chance to watch them. Somehow, I think there must be a superior amongst these minnows—a commander, like a queen in a swarm of bees, or the leader of a colony of migrating ants. Perhaps in this case it is a king, it is hard to say. What is certain, is that the gathering of the clan is always started by a limited number, who travel about collecting, gradually, what becomes an army.

I have seen the initial band pause by a big stone until twenty or thirty others of all sizes have come out from beneath it. The same thing has been repeated, by the ever-increasing horde, time and again, up and down the stream, until the assembling is complete, and the numbers swelled from scores to perhaps many thousands. The time of year has come for them to move. The urge to feed is prevailing, and they know sufficient food is not to be obtained in the sidestreams for such a mighty throng.

With what appears to be one general impulse they all move from spring stream to river. And after they have gone you can search the stream, peer beneath the stones, under the banks and delve in the mud where, but a few days previously, there were minnows concealed everywhere now not a single one is to be found. They have gone to places where, in the higher temperature water of the main river, their food is daily becoming more plentiful, and where the gathering for spawning will take place.

Early one August, I came across one of our fishermen busily casting to some trout in one of the shallow reaches. I watched for a long time before interrupting him and making my presence known, as I was anxious to see if he could indeed catch one of the many good fish that were lying within easy casting distance of him. Time and again he threw the dry-fly he was using, but without result, and then, apparently in desperation, he changed to a nymph and repeated his efforts with that.

Still he had no success, yet fish could plainly be seen as they rushed here and there over the gravel bottom. Every now and then the water was broken or distorted by waves and bulges, whilst a tail or dorsal fin showed for a moment above the surface. He jumped as I spoke, for so intent was he in trying to catch one of the fish that he had failed to even sense my approach.

"Hello, Sawyer," he greeted me. "You're just the man I wanted to see. Just what are these fish taking? I've been here for hours and they keep on moving to something. Though I've tried everything I have, not one has shown the slightest interest. Look at that big fellow. Now, surely he's nymphing."

Well, as I told him, I would have been most surprised had he caught one of the fish with fly or nymph, for these fish had congregated where the water ran swiftly over the bright gravel-bed and they were feeding on minnows. Indeed, had he been more observant he could not have failed to see the milling hordes of bright-coloured little fish as they swarmed over the bottom, for they were spawning. Indeed some were flicking up and out of the water, and others were scurrying here and there with total disregard for the dozen or so big trout that were lying amongst them.

I suggested to the fisherman that he should draw nearer to the edge of the river, put down his rod and watch. To this he agreed. Some of the trout were over two pounds, and at a glance it was obvious that they were in excellent condition. One big fellow appeared to be the boss. He had taken up a position just to one side of a big stone. But he was gorged and less active than some of the others. Minnows swarmed over his head, tail, and around his body, yet every now and then, though full, he was goaded into activity, and with a sudden rush with mouth agape, he would surge into the masses and snap a luckless minnow. Occasionally he would chase away another who had ventured on his territory. As each time he, or one of the others, made a slash at the minnows, so they scattered in all directions as though with one general impulse, while the surface of the water was distorted by the movement.

The trout varied greatly in coloration and my companion was much interested in three which had dark bars on their sides in a manner somewhat similar to the colouring of a perch. But I had noticed this colouring many times in past years, and it would seem that it can be brought about by trout continually feeding on a fish diet, especially minnows. One of the fish we were then watching had been under my frequent observation for over a week. When first I saw him, he was very light coloured and had a very noticeable injury to the dorsal fin. The change in colour had been gradual, but it had definitely altered, and now the dark-barred effect was quite pronounced. It could be a form of camouflage, but this I feel is doubtful, otherwise, surely, if the pigmentation could be altered as an aid to the fish in getting their prey, all of the trout then present would be of the same colour.

The colouring of perch is well known and these feed mostly on minnows and other small fry, but I have seen pike, especially small jack, with this dark-barred effect in their pigmentation, and in the instances when I have examined the contents of their stomachs the fish have been feeding on minnows. My companion had had ample proof that when trout are feeding on spawning minnows it is difficult to interest them in any of the legitimate patterns of fly or nymph. I felt sure, when I suggested he should watch, that he would be interested. But he was more than interested, he was fascinated, and when the time came for me to go I left him there, still peering into the water, with his forgotten rod propped upon the bank behind him.

Usually, later in the year, the great shoals of minnows present in the Upper Avon again congregate into vast assemblies which sometimes amount to hundreds of thousands to begin their migration to the warm waters of the spring-fed sidestreams.

During the last week in September of 1949 I saw one such multitude on their way up-stream, and the circumstances of their migration interested me so much that I made a note of it. They had run up a reach of water, and all were gathered in a large pool down-stream of a set of sluices. One sluice was partly open and through this the impounded water from the above reach was rushing in a torrent of such swiftness that it created an impassable barrier for the little fish. I watched many of them make unsuccessful runs in the fast current, but they were quickly washed back into the pool.

At one end of the set of sluice gates was what we call a half-hatch, an arrangement by which the top half of a sluice gate can be drawn to allow water to pass over the lower part, and so maintain a certain level in the reach above. When open, this half-hatch creates a miniature waterfall from one level to the other—a fall of about two feet. I had no thought that I should assist the minnows when I opened this half-hatch to reduce the impounded water up-stream, but within a few moments of my doing it the little fish took advantage of a means by which they could gain the upper level. The water was perfectly clear, and I watched the great shoal milling round and round the pool. Then, suddenly, they formed off into a long file and came towards the waterfall. Minnows of all sizes were present, but those in the lead were fully-grown fish of between 3 in. and 4 in. in length. With a speed that surprised me they approached the falling water.

I watched for a long time. I was fascinated to see the

little silvery bodies flipping up and out of the water, sometimes as high as a foot or eighteen inches. Some ran successfully up the fall into the quiet water above, but many others
dropped in a continuous shower on either side of the fall,
back to the lower level. As they fell back others took their
places, for just down-stream of the surging water a milling
mass of lively little fish waited their chance to try their luck.
But after a while a great mass of flannel weed came through
the open sluice and swirled into a position just to one side
of the fall. There it jammed fast in the rushing water. For a
few moments the water was dirty and the minnows stopped
running. Soon they started again with even greater concen-
tration than before. However, they now made their run
from a different angle, possibly the current had been
diverted, for instead of swimming from immediately down-
stream of the fall, as they had been doing previously, they
now approached from the side opposite to the bulk of
weed.

Above the falling water the fish leapt, and fell like a rain
of silvery shapes, but now misfortune awaited them. For as
they hurled themselves clear, so many of them fell on the
bulk of weed, and there struggled unsuccessfully to regain
the water. Soon the weed pile was a mass of glittering
bodies—a mass which continually received others. The
weed was indeed a death trap.

Hordes of minnows are a pest in a trout stream, for they
eat many of the tiny organisms that could be a food supply
for trout fry. They also make heavy inroads into the stock
of immature fly life. So, for the time being, I was pleased to
see so many of them hurling themselves to destruction. I
continued to watch whilst the shoal persevered in the desire
to get up-stream. I saw many which, time after time, made
the attempt to run the fall. I saw some slither off the edge of
the weed mass and at once make another try, and so struck
was I with their relentless purpose that I was moved to pity
those that were dying in the attempt.

Perhaps I am a sentimental fool at times, I do not know.
But no longer could I sit and watch them die in an effort to

join their brethren. I quickly got a rake and worked until I had cleared the mass of flannel weed away from the danger zone. Some of the minnows stranded on it were still alive, and gathering them all into a big tin I took them all to the upper level and turned them loose. With a queer lightness of heart I saw them swim away in the deep water, and then I sat down to watch a wonder of Nature, unmarred by tragedy.

Somewhere, scattered about in the river bed and banks—in the sidestreams, ditches and drowning carriers, in ponds, brooks and spring heads—hundreds of eels wait in due season for nature to provide them with an easy and safe passage to the sea. Sometimes such a chance will come in October; in other years the eels have to wait until November or even December before conditions are such that they can bid good-bye for ever to the waters of the Upper Avon. Such eels are mature and have spent perhaps ten years in fresh water. Nature asserts her influence and to them a great urge comes to start their run to the sea, and (according to scientists) to commence a great journey of thousands of miles that will take them to their spawning grounds in the deep waters of the Atlantic Ocean. Eels are queer creatures. I have studied their habits since I was a boy, but each year I find out something about them that sets me another puzzle to solve. There is something sinister and mysterious about these creatures of darkness, and their habits have been a continual controversial subject for hundreds of years. I think of them as wanderers, the nomads of the river, for, from the time they enter the river as tiny elvers, until they reach maturity, their lives are spent in ceaseless movement from place to place in search of food.

Each year is but a repetition of a former one, just a continuation of a great cycle that has been going on through the ages. Young eels come into the river in the spring, and mature eels leave in the autumn, and so it will go on. I often stand on one of the old bridges which span the river, and think of the thousands of eels that, at some time, have passed beneath its arches. What a story the old stonework could

149

tell, for hundreds of slimy bodies have moved along the supporting piers in passing up-stream or down.

Each autumn other wriggling bodies will pass beneath this same old bridge, making an attempt to return to the place of their birth. But I am doubtful if anyone will see these migrants as they move along on their journey, for when this happens it will be on a dark and stormy night, with the wind whistling in the withy beds and rain hissing amongst the sedges on the river banks. And the dirty water of a rising river will be streaming swiftly between the weed-beds and smelling of freshly moved mud and vegetable matter. Perhaps lightning flashes will, from time to time, show the autumn leaves being stripped from the trees, while claps of thunder send tremors through the ground, as they rattle over the downs like a barrage of artillery.

No night this, for humans to stand on a bridge peering into the water, but it is these conditions that the mature eels await. To them such a night is ideal, for, cloaked in the darkness and dirty water, they are safe from all their enemies and in the fast-flowing river they need expend little energy to travel down-stream. So these nomads leave the river, never to return. No doubt there are many creatures in the river who will be pleased to know this seasonal migration has taken place. Though there will still be other immature eels to harass them, a menace has departed, for during their life in fresh water many thousands of smaller animals have died to bring these eels to maturity.

Throughout the years much fiction has been written about rivers and river life, and when facts are advanced they are apt to be treated with suspicion. Years ago, someone started romancing with a pen and the views expressed have so often been repeated by others that they have come to be regarded as authentic. Give a dog a bad name and it remains throughout his life, but fortunately dies with him. Give a fish a shady record and it is passed on and on, from generation to generation, until something is done to disprove it.

There has been a lot of controversy about otters on the

question of whether they do harm or good in a trout stream. Though it is generally admitted that these animals do kill trout and salmon, it would seem to be considered that, because the otter will kill and eat eels, he vindicates himself for the comparatively few game fish he is said to take.

Nothing good has been said of the eel, yet is he as big a villain as people like to make out? He is said to be a ruthless destroyer of trout eggs, alevins, fry and even yearlings. I wonder on what these observations have been based. Can anyone truthfully say they have seen eels routing into a trout redd for eggs, or for alevins? Has anyone ever caught an eel and found trout eggs in its stomach? In fact, have they ever caught many eels during the spawning time of trout? I write of events as they occur in Nature, not of what happens artificially.

In the south country trout spawn in January and February and during these months, unless it is is an exceptionally mild winter, eels are hibernating. They are deep in the mud and silt of the river bed, tucked away in the banks, in the rush-beds, and other such places—well content to remain curled up until the rising temperature of the water warns them that spring has arrived and with it the food they desire. They have no need of food, least of all in the chilled water of winter, to cause them to go routing in gravel for trout eggs and alevins.

People who have seen naturally bred trout fry will know of their habits and the scant likelihood an eel has of taking more than an occasional victim. As for yearlings, where eels are concerned I think these little trout are quite capable of looking after themselves, though perhaps now and then one may be taken unawares. A river might be better off without eels in it, but even as the otter, the heron, and other mistakenly persecuted creatures, they do certain good. Eels are fond of loaches and of minnows, and take large numbers of both. They will also feed on the eggs and offspring of coarse fish—of roach, dace, pike and minnows—which are available during the feeding time of the eel. I have often caught eels when they have been lying amidst a horde of spawning minnows and have found as much as half an eggcupful of minnow spawn in them, as well as mature fish.

What of the harm all these coarse fish could do had not they been destroyed by the eels? Think of the thousands of immature fly larvæ that coarse fish and minnows will consume, as well as of the damage that pike can do amongst trout, and then think twice before condemning the eel as a worthless scoundrel.

One often hears of fish in isolated ponds and how at times they are discovered when the water has dropped to a low level, or has been entirely absorbed during a season of drought. This brings to mind a question which from time to time has caused some controversy—"Can eels travel over dry land?" I have spent most of my life in a river valley, and my father and grandfather also had much to do with water, but so far I have never seen an eel making an intentional journey over dry land, and neither my father nor my grandfather could tell me of ever having seen such an occurrence except in a water meadow. Though there are many people who have written to say they have seen eels travelling overland from one place to another, I am inclined to be rather sceptical and to think a wrong conclusion has been placed on the appearance of the creatures in such unusual circumstances. I have never yet read of anyone seeing an eel, or eels, *leaving a pond*. In most cases the eels have been discovered *en route* midway between two respective waters. There has been no proof put forward that the creature has indeed been prompted to move from one water to another.

I read an interesting account where someone found an eel caught in a rabbit trap that was set between two separate waters, but this bare fact proves very little. I once saw an eel travelling along a village street. This one had been dropped by a heron which someone had frightened as it was passing overhead. The eel was not badly injured, and it is possible, it being raining and a certain amount of water in the gutter, that it would eventually have entered the river once more, as it was heading in the right direction. An innocent observer

152

might have thought he was witnessing a remarkable case of migration.

However, this is by the way. What I would like to know is, if eels can travel overland (as it is claimed they can) how it is that many hundreds will *choose to die* rather than move away from an isolated pond in time of drought? Here they have every opportunity to move if they wish, indeed, they have a dire necessity to do so, but they do not do so. They stay in the pond while the water level sinks lower and lower, until they are choked in the puddle of muck that is left in the lowest part of the pool. They wait, with death facing them, for water to fill the pond once more to over-flowing, so that they can travel as Nature intended they should, but many have waited in vain.

I cannot think that eels can travel when and where they will. Many times I have thrown an eel out of the river on to the dry grass of a meadow. There it has squirmed this way and that in an endeavour to move along, back to its native environment, but without success. An eel cannot move very far unless the slime and scales which cover its body are kept wet. The eel is known as a slippery customer, but just cover one with dry dust and it is quite easy to pick it up. In fact it is rendered helpless. I think the tales of eels travelling overland originated with some of the old drowners who tended the water meadows. Their stories were perfectly true. Undoubtedly these men saw eels moving at times overland, from one drowning carrier to another, from one runnel to its neighbour, or to a drain. But in moving about the meadows the eels were in wet grass, on wet land, where their progress was not impeded by lack of moisture. In fact, the conditions in a meadow are very little different from the brooks and rivulets of a valley, where eels travel at will.

Except during the spawning season, the river lamprey, or sand pride as it is often called, has little value in a trout stream, yet I have spent many an interesting half hour watching them. These little eel-like creatures belong to the fish family and seldom exceed six inches in length. They spend most of their time concealed in the mud and silt of the river bed and banks, where they find food. Often they are mistaken for small eels, but the formation of the sucker-like mouth, and an absence of pectoral fins, is sufficient to tell them apart. Once I sat watching a group of eight lampreys. Their spawning season often starts about the end of February and for this purpose they leave the silt and mud and go to a place where the bed of the river or stream is clean and gravelly. These eight were busy with their spawning.

It was in a sidestream, where clear spring water, about nine inches deep, flowed gently over the clean gravel bed. The site chosen was near a large stone firmly embedded in the bottom. Some years ago I determined the sexes of sand prides, so that I now know them in the water, and can interpret their activities. The slightly lighter colouring of the females, together with their habits while spawning, were sufficient to tell me that of the eight only two were of this sex. The actual shedding of the eggs and fertilisation is little different from the process carried out by trout or grayling, excepting that both male and female attach themselves by means of their sucker mouths to stones or gravel on the river bed, and there make the flexious movements which extrude eggs and sperm. But what interests me most is the work and showing off that is done by the males.

For it would seem that each male vies with another in bringing a piece of gravel to the nest, and, if this pleases the lady owing to its size, or possibly formation, then he is allowed the place of honour. On being presented with the offering, the female will leave the piece to which she is attached and join the male in sucking at the stone he has brought, and together they will spawn. Pieces of gravel, up to the size of a large garden pea, are brought from a radius of twelve or more inches. I have seen these little worm-fish trying to struggle along with stones much larger, indeed they have moved them for several inches before abandoning them.

All the while the lady remains near the nest, either

spawning, or arranging the bits of stone to her satisfaction. For it would appear that these little pieces of gravel are a means by which she tries to conceal her eggs, and at the same time she sets them so that the current will pass under, round, or over them, and thus help the incubation.

As I watched, I saw these six males each bring an average of half a dozen stones to the site. I saw the same stone taken away and brought back three times. Possibly it was by the same male, who knew he had something good, but of this I could not be sure. Many of the offerings were rejected by the two females as unsuitable, and these were carried a few inches away and abandoned. The strength of these little creatures is remarkable, for as well as being able to attach their sucker mouths to, and carry, big stones, the threshing, flexious movement of tail and body while spawning is sufficient to move stones as big as a walnut. Sand prides spawn in shallow and deep water alike. I have seen them in five inches, and on the river bed beneath five feet of water. During their activities they are often taken by pike—both small and large. I once caught a pike with twenty-three inside it. Many of these were still alive, and indeed swam away when I returned them to the water. Trout, especially spent females, will also take them from their spawning grounds, and I have found them in herons and dabchicks. So for a short while they provide food for both trout and the enemies of trout.

Strange, is it not, how a name learned in boyhood days will remain in memory all one's life? Jimmy Goblin was the name by which, as boys, we knew and spoke of the bullhead, miller's thumb, or, to give it its Latin name, *Cottus gobio*. The words goblin and *gobio* have a certain similarity, but where the Jimmy part came from I cannot say. We used to have great fun in the summer holidays hunting for these queer-looking creatures beneath the big stones, and in chasing them as they shuffled from place to place about the river bed. However, this is by the way. I now know the fish as a bullhead and personally think the appearance of the creature truly portrays its ugly habits.

155

In a trout stream the bullhead must be considered as a pest, as it will take trout eggs, alevins and fry. But it also does some good, so it cannot be condemned out of hand. It is quite possible that if all its habits were disclosed by research we would find out for certain just why it was included by Nature to be an occupant of river and stream. My own observations give me reason to think it is just one of the many creatures whose sole purpose in life is to maintain the balance of Nature. Though I have found them in the stomachs of trout, pike and grayling, it has been the exception rather than the rule, and I think the bony and spiny formation of the bullhead gives it a certain immunity from larger fish.

Most of their life is spent in the semi-darkness beneath cover on the bed of stream or river, and here it is that they spawn during the months of April and May. The shape of the head and the bony structure, combined with the spine—armed fins, enables the bullhead to force its way into the gravel of the river bed and to move beneath boulders and big stones. Though it feeds on a great variety of small creatures, I think the principal work of the bullhead is to kill off and eat all the weaklings of other fish species—the unhealthy small alevins and fry, which, being delicate and sickly, have taken cover beneath stones and other debris. The mouth and stomach of a fully-grown bullhead is sufficiently capacious to allow it to take minnows of large size. I once caught one which measured $3\frac{1}{2}$ in. in length just after it had partially swallowed a minnow that was 2 in. long. More than half of the tail end of the minnow was still protruding from the mouth of the bullhead. This was during the winter (January, I think) when minnows were hibernating in places where the bullheads make their homes.

The bullhead's method of egg-laying is interesting and has some resemblance to the procedure carried out by sticklebacks. The eggs are laid in clusters of perhaps two or three dozen and are attached to the undersides of stones, where a current of water can pass around them to provide oxygen. How the female accomplishes this task I cannot say, for it would seem that at times she would have to be upside down. In comparison, the eggs are larger than those of any other fish I know, but I cannot say how long it takes them to hatch. Frequently, I have found them when eyed, and have been amazed to see the very high percentage of fertile eggs that are in each cluster.

The eyed eggs are interesting to watch and many times I have been amused to see the evolutions of the tiny fish as they try to break the shells of the eggs to hatch. Both eyes are then plainly visible and it appears as though the eyes are unattached and are whirling about inside the egg-shell. The freshly hatched bullhead is a queer creature, for, with its big head and ungainly yolk sac, it has far more resemblance to a lump of eyed jelly than to a fish. It would appear that, like the stone loach in whose company the bullhead can frequently be found, the bullhead can only live in well aerated conditions. At times I have tried to breed them in static water but without success. Like the eel (another forager) the bullhead is a wanderer. His life is spent in moving from one big stone to another, and possibly it is then that occasionally one falls victim to a larger fish.

A stone loach is an excellent bait for pike, but those of suitable size are not always easy to obtain when they are most needed in the winter months. Corrugated iron sheets can help. They are favourite sheltering places. I suspect this is because the corrugations trap and create a steady flow beneath them, and so keeps the gravel free of silt and mud. I find it a good plan to encourage loaches to use these iron coverings, and often introduce a few of suitable size for easy handling into some of the clear, shallow streams that I know. In this way loaches can be localised, and when some are wanted for baits then it is quite a simple matter to find them. All one has to do is to locate the coverings, lift them up and select any of the fish that are of the right size.

It all sounds so easy, but to take loach from a river or stream-bed is no simple matter. They are not swimming fish. They move about by shuffling along as close to the river bed as they can, so that it is very difficult to get them to enter a net. They are not easily alarmed, but when they are, can move with a speed that is surprising, often to disappear as though swallowed into the very bed of the stream. I use considerable caution when lifting up these sheets of iron, and so far have found nothing better to catch them than to use a bucket that has no bottom to it. I know it is rather a primitive thing to go fishing with, but often a simple device will beat methods that are elaborate. The plan I use is to lift the sheet slowly from the side parallel to the current, gradually bring it to the surface until I can see beneath it. Slowness in moving the sheet is essential or currents will sweep the stream-bed and cloud the water. Then, when suitable-sized fish are located, place the bucket over them and jam the rim well down into the gravel. The loaches are then trapped. All that is then necessary is to grope around in the bucket with your hands, until they can be grasped.

Loaches have a wonderful camouflage. Their colouring merges so well with their surroundings that often a very close scrutiny is needed to detect them even though they may be lying without cover of any kind over them. I have used the bucket system for catching little fish (and I must confess large ones also) since I was a boy. My father taught me the trick and it often comes in useful.

12   THE ARCH ENEMY

PIKE were present in the water where I am keeper long before I was born, and I fear that they will still be a curse long after I have ceased to harass them. To keep their numbers in check is a never-ending battle but it is work that I consider to be the foremost of my occupations.

Nothing saddens me more than to see good trout water ruined, and twice recently I have had a look at lengths of the Avon, which, in years before the last war, held a good stock of trout, and provided excellent fly fishing. In both these lengths the stock of trout had taken on another form, for they had gone to build up the bodies of the long stream-lined brutes that were in possession of every pool and eddy. It is nice to know that most of the tyrants are dead, for just lately the owners of both lengths have had their waters netted. But the damage has been done. The killing of these hordes of pike will not put back into the water the trout they have eaten, and ahead lie years of re-stocking.

In one of these lengths, a reach of about two and a half miles, just over one hundred pike, many of them between 6 lb. and 11 lb., were killed, but all of them had long since finished feeding on trout, and when caught, were long, emaciated creatures which should have weighed much more. Indeed, I think they were half-starved, for in

159

addition to clearing almost every trout in both deep and shallow, they had turned their attentions to the grayling and coarse fish, so that few were left in the water. I think also that they had been feeding on their smaller brethren, for few small pike were seen or netted. These pike were caught just in time, for up-stream there are still good stocks of trout, and I feel sure that many of the larger pike would have migrated with the first winter flood, and soon would have made heavy inroads to the stock of trout of another fishery.

In past years, this same water had held upwards of 1500 trout of all sizes to the mile; in fact one had no difficulty, on a nice summer day, in counting a couple of hundred fair-sized fish as one walked along a bank, and at times over 400 trout of from 1 lb. to 2 lb. or more had been killed there by fly fishermen in a season. But at this time it was netted annually, and observation kept throughout the year to prevent pike from getting established. It was ten years since it was last netted. I did it, and I knew there were still small pike in the water afterwards. It was impossible to get them all.

During this ten years, the fishery had run wild. River and bank vegetation was allowed to flourish unchecked. Sheltered by river weeds during summer and autumn, and by the clouded water of winter, the pike thrived unnoticed, and caused little anxiety until it was too late. I once knew an owner who said he considered it to be worth £1 to kill a pike in a trout fishery. He changed his mind when I killed a dozen in his water one day, and I got a pint of beer instead. But this is by the by. Had he said they were worth £10 to kill, he might be nearer the mark at the present value of trout. It is safe to estimate that between them the hundred pike in this two and a half mile length had destroyed £3000 worth of trout in ten years. As I have said, this water frequently produced an annual tally of over 400 trout averaging 1¼ lb., while some 2000 fish of all sizes remained as a stock for future seasons.

I fear with pike it is a case of "what the eye doesn't see the heart won't grieve over." If only one could be present and witness the death of every trout, or even to see some evidence of their having been killed, than I feel sure far more attention would be given to ridding our rivers of trout's worst enemy. Many people deplore the fact that an otter has taken a trout and left half of it on the bank, or go out of their way to kill a heron because they have seen him with a fish on the river-side, yet there, perhaps within ten feet, lurking in some weed-bed, or some deep pool, is a menace that is always present and depleting the stock day and night. Just a pike they say, yet here is an enemy from which trout have no escape; who, if he misses one to-day, will surely take his toll to-morrow. If trout are to live and thrive, then the human must help them, for apart from the otter, a pike of over 1 lb. in weight has no enemies excepting his own kind.

I saw the pike of which I have written. In fact, I killed most of them as I was in charge of the netting. But it gave me no pleasure to see the row of vicious bodies lying side by side. The sight was poor compensation for the memories of years past, when on the gravel bottom and behind each weed-bed, the bright-spotted trout had added to my joy of a walk along the riverside. Few people fully realise that to have pike and trout together in a river is no different from trying to keep foxes and poultry in the same run. Who can blame the pike if they have a preference for good food.

I have been told that I must have a sadistic streak which is manifest in my dealings with pike and I must admit that it does give me immense satisfaction to be able to administer the last rites to the brutes whether they are babies or grand-fathers. Each time I end the life of a pike I think of the life he has taken and I picture myself as an appointed executioner.

Many and varied are the methods I use in the destruction of *Esox*, for between us there must ever be war. He is a worthy opponent and he has Nature to help him. For, though during the summer months his cloak of security is sometimes stripped from him by means of weed-cutting and clear water, he has a great advantage when the high cloudy water of winter screens him from the keenest eyes.

It is then I use a rod and bait and fish for him in the time-honoured way. From October to April I periodically spend an afternoon trolling. I am not in favour of live baiting for pike; I think, if it is necessary that anything should suffer inconvenience, then let it be the pike—he has a brutal nature and it is only fitting that he should get a taste of his own medicine. No matter how well a live bait is arranged the little fish is bound to suffer, either in body through being attached to hooks, or in mind through being tethered and at the mercy of its greatest enemy. Without question live baiting is attractive but to me, attractive only to the pike. I have no desire to stand for hours on a bank watching the struggles of an active bait, glorying in the fact that the more it struggles and shows itself, the more likely it is to attract a pike. For though this live baiting is done universally, one can still catch pike and, what is more, catch more pike and have greater sport by using a dead bait and one's own skill to make the dead appear to the pike as though it were alive.

Such a bait I use mounted in a manner somewhat similar to that adopted and recommended by Nobbs in the early nineteenth century, but instead of having hooks sticking out from the mouth of the bait, as was his custom, I seal the jaws with thread and attach a fair-sized treble hook to the trace near the tail of the bait. For baits, I find a big stone loach—a 4 in. to 5 in. trout, or a half-grown frog—to be the most attractive to big fish and small ones alike. These small baits are easy to mount and to control in the water with light tackle.

Fished through the pikey holes in a sink and draw movement they prove an attraction few feeding pike can resist. The pike may take whether it is being drawn to the surface with a lifelike wobble or whether it is sinking (and my method is to weight it with lead in the mouth so that it sinks head first). He quickly turns the bait so that it slides head first towards his throat and the hook on the trace comes into his jaws. A pause of a few moments with slack line is sufficient to allow the pike time to get the hook inside his mouth, then a firm but gentle lift of the rod will drive home the barbs. If little pike have to be returned to any water where there is a size limit, it is quite a simple matter, provided a gag and disgorger are used, to remove the one treble hook without damaging either bait or victim. I have many times taken up to half a dozen pike of varying sizes with the same bait. To some, it might seem a shame to sacrifice a little trout to make a bait, but in trout streams I consider the sacrifice to be well worth while. The death of one little trout can save the lives of hundreds of others and, whether large or small, what more noble ending could a trout have.

As everyone knows, in catching *Esox* one needs luck as well as skill. There are days when no matter what bait you use, or seemingly, how you cast it, a pike is ready to take and be landed. Often enough one good day will follow another; you may perhaps get a week when pike are really feeding. I went out on a Friday afternoon in mid-December a few years ago. The river was low and clear for the time of year and pike were well on the move. I got several up to about 4 lb. and decided I would try a lower beat the next afternoon, where there was a faint hope of getting something larger.

In the village I had a friend who often gave me a hand with weed-cutting, netting and various other jobs about the river. He was a good hand at catching pike with wire and harpoon, but had never used a rod and bait. Knowing he was keen to try, I fixed up a second rod and tackle, and together we went to the lower beat. Even if he did not catch a fish he would be company, and I like a companion while pike fishing. Each of us had a big stone loach mounted, to troll head first in the sink and draw method I find so deadly. I gave him a short lesson in casting and in working the bait, and then he took one bank and I the other, to fish our way down-stream.

Pike were not thick in this water, and those on the feed seemed to be all under my bank of the river. I got four fish up to 3 lb. as we worked down-stream a half mile, but my friend had not moved one. Then we came to a place where the river course made a big S bend and there is a deep pool in either bend, both a favourite place for pike. I fished the up-stream pool, my friend the lower one where there was a good area of very quiet water. Almost at the same moment we both ran a fish, and following my lead, my friend slipped the clutch of his reel and stuck his rod into the bank, to wait until the fish had turned the bait and started to move off with it inside his mouth. I watched my line jiggling where it entered the water, and could imagine the pike sliding the loach across its jaws to take it head first, and then the line moved up-stream. I lifted the rod and tightened. Apparently, my friend's fish moved just then, I saw him grab his rod and give a vicious strike with it. Then by the bend of his rod I could see he was fast in a fish. Mine gave little trouble, and soon a 5-pounder ended his life of eating trout.

Dropping rod and other gear and taking my gaff, I ran down-stream to be opposite my friend and, if necessary, to give him some coaching in playing out the fish and gaffing it. By the bend of the rod I could see he had on a good pike, and it was playing deep down. Ten minutes passed. All we could see was the line where it entered the water, first moving up-stream twenty yards or so, then down to the tail of the pool, a similar distance below. I did not think he had on anything exceptional, and thinking the bend of the rod was an exaggeration owing to its age I told him to hold tighter and bring the fish to the surface. To me it had appeared as though he was just letting the pike run where it chose to go. He reeled in as the fish came towards the rod in one of its runs up-stream and then put on pressure to bring it into sight. The moment I saw its size I knew I had made a mistake, but it was too late. The pike and my friend saw each other at the same moment and both had the same reaction—one to get away from the other. As my friend went backwards with astonishment, the rod bent to a half circle. What I feared happened at that moment, the pike gave a lunge and the rod broke in two pieces just above the

cork grip. He was left with 2 ft. of handle and the reel
spinning like mad. But even as the top part of the rod was
sliding down the line and into the river, he grabbed it.
There he stood, not knowing what to do, reel and butt in
one hand, and the rest of the rod in the other. I could see
the pike was still on and shouting to him to keep a tight line,
I ran down to where I could wade across a shallow, one
hundred yards down-stream.

As I reached him the pike gave another long run. He had
been trying to hand line the fish, and the reel jammed with
line around the handles. Never could a fish have been more
firmly hooked than that one, something more than luck
was playing a part. It came to an abrupt halt which pulled
the butt part of the rod tight against the lower rod ring of
the other part, and I scrambled to get the line free as the
big brute jumped clear of the water. But the fish had turned
once more and was coming up-stream. For a moment I was
afraid it had escaped. Grabbing the butt, I started to reel
in line while my friend held on to the other part of the
rod to feel if the pike was still on. It was. And in this
manner we continued the fight, he holding the 7 ft. or
so of the rod top and I the butt and reel. Between us we
played the fish up and down the pool trying to get it near
enough to gaff. My gaff was lying by my side as I crouched
at the edge of the pool ready for the first opportunity, but
it was a long time in coming. Time and again I was afraid
the fish had escaped, for my friend had little idea how to
bring such a one to bank, and kept frightening it. Two or
three times the big brute sailed past in full view, when a
little more rod pressure would have brought it within reach.
Then came the chance for which I had so patiently waited.
I drove home the gaff and out he came.

That fish weighed just 15 lb., and though I have caught
others by less sporting methods and with less excitement,
which weighed up to 20 lb., it still holds the record
on the Upper Avon for a pike caught by rod and bait.
I have had two since then which weighed only a few
ounces short, but still I have to admit that at a first

attempt a pupil has beaten the master. I often wonder how things would have fared had I not been there and, after all, did he, or did I, get that one?

Time and again during November and December, when hunting for loaches for pike baits, I have gently lifted a big stone or some other debris and there where it has rested, instead of the loach I have expected, has been a big frog crouched in a hollow in the gravel. There he sits, merging with the general tone of the stream-bed with not a movement of the body or of muscle to betray him. He looks through the clear water with a kind of reproachful stare, as if to say: "Well, and what do you want? Why can't you leave a chap in peace to enjoy his rest for a while?" It makes me feel that indeed I am taking a somewhat mean advantage and so I gently drop the stone back into place once more and leave him to his seclusion.

It is difficult to know if the frog was meant to be a creature of land or of water, the land provides his food supply and the water his security. Usually it is early November when the mature frogs leave the land for the water, and in the water they then stay until after they have spawned. As a boy I used to wonder where all the frogs came from when they gathered together for spawning in January, and had the idea that they had just migrated from the land. What puzzled me was the fact that I never saw frogs anywhere but in the water and of course the reason was that the creatures were already there—that they had been hibernating in the streams for the past few weeks—just resting and waiting the call of Nature.

Until spawning takes place the frog is an individualist. I seldom see more than one in the same place during the hibernation period, and, in November, when they go to the water they do so as individuals. Each frog selects his site for retirement and there he sits in solitude until prompted to move to the spawning-beds. Most of them prefer water that is near to a springhead. This possibly is because spring water maintains a fairly even temperature and during very cold weather is much warmer than static water or that exposed in a river. Spring water near a spring-head never freezes.

What has also puzzled me is that the frogs one finds in the water during the winter months are always fully grown. How long it takes a frog to reach an age for spawning I cannot say, but obviously it is more than the few months between May and October and, thinking it over, it would appear that it is only the fully grown ones, those expecting to spawn, that go to the water at the beginning of winter. I have seen quite small frogs in October, but so far have not discovered their winter quarters.

Their life in the water is not without danger. Some are taken by pike and by trout. Chub are also fond of frogs. But the chief enemies are the otter and the heron, and both of these are well acquainted with the frog's hiding places. Often I see where one, or both, of these predators has hunted a spring-fed stream. Then I find where otters have routed beneath or turned over large stones, or where a heron has stalked through a bed of watercress or celery, and there on the banks is plenty of evidence that they have been successful. Both otters and herons are very fond of frogs and will often take them in preference to fish.

I cannot say that I like big frogs, but little ones always attract me. There is something so tiny and compact about a little frog that when they have lost their tadpole tails and I see them about the meadows as they are migrating in May or early June, I feel, though I have done it scores of times, that I must pick up some and have a closer look at them. A migration of tiny frogs always interests me. Usually they wait for a storm of rain so that the moisture will aid them in their movements, then for a little while every part of the valley seems to be alive with tiny hopping and crawling bodies. For a while their life in the water is over. Together they move to fresh homes and food in the surrounding countryside, there to await the time when, fully-grown, they can return to the springheads.

Though one may walk a dozen times along a reach of water looking for pike, the fact that none are seen is no reason to assume such fish are absent, for Nature has provided *Esox* with a method of camouflage of which he makes full use, to screen himself from the keenest of human eyes. Even though it might be easy to see into the water and to search the river bed in the likely places, a very close scrutiny is necessary, for unless some movement is made by the fish to attract attention, a considerable number can be passed without being seen.

Often, even though a view of the actual fish is not immediately obtained, the movement made by a pike when striking out from a bank is sufficient to tell one of its presence. They have a movement in the water that is characteristic of the species, and which can betray them to anyone who recognises it. Pike often lie close under one or other of the banks of a river. They have no liking for moving water, and frequently there is a quiet backwater, or an eddy, where they can rest in ease. But even though one may fail to see them lying in such places they are often disturbed by movement along the bank. They become alarmed, and strike towards mid-stream or to deeper water.

In this striking movement, a pike uses all its ventral fins in unity with dorsal and tail in making a synchronised thrust to propel its streamlined body through the water. The thrust is often so great that it carries it at speed to mid-stream, or to a point several yards away, without further movement of fins. The vacuum caused by the thrusting fins usually sets up a swirl in the water, which disturbs the river bed, and forms a cloud of dirty water just as though an explosion has taken place. This swirl of dirty water, coupled with the straight path of the wave the moving body creates, leaves a story easy to read.

Though pike may strike out from the bank, the chances are that they will lie at the point to which their initial thrust has carried them. I think they are curious to know just what has disturbed them and, if caution is then used, this curiosity may lead to their being easy victims to wire or harpoon. Should they be alarmed a second time, then the chances are that they are fully aware of the nature of the threat and will not be easy to catch. Usually, the wave made by a moving pike is quite easy to see in shallow water, but a pike of several pounds can cause a wave to appear in water five feet or six feet deep. Such a movement of the water is easy to follow and if one watches the wave until it stops moving, the pike can usually be discovered lying on the river bed within a few feet.

I think the scales of pike have a mirror effect, in that they catch and reflect the surroundings in certain lights, for, occasionally, no matter what the size, or where they may be, they seem to merge with the river bed in a manner that is uncanny. But at the base of all the ventral fins there is a lighter part that always remains the same, wherever the fish may be, and it is for these parts that I look. Once these light parts have been seen it is surprising how quickly the remainder of the fish comes into view. I have known my assistants to walk half a dozen times past pike that have weighed 10 lb. to 15 lb., and these fish have been lying on the bottom without cover of any kind over or near them, and in water clear enough to see every stone on the bottom. I have passed them myself, and yet when discovering them a few moments later I have wondered if I had been blind in my previous scrutiny.

A friend of mine was telling me of an experience he had just previously when he was spinning for pike. Apparently, as he had been walking along the bank close to the river's edge, he had seen what he thought was a dirty branch of a tree lying on the bottom amongst the roots of some rushes. Being suspicious he bent to get a closer view and startled a moorhen which scuttled through the rushes. The "log of wood" then materialised into a huge pike which struck towards the centre of the river amidst a flurry of muddy water. He told me he tried for an hour afterwards to attract the great brute with his various spinning baits, but had no response. I was not a bit surprised, for what he had seen I have seen many times before, and I know the folly of trying to interest such a fish in any kind of bait. For more than a week the river had been in flood, a fast torrent of muddy water following several days of very heavy rain. The pike he had seen was doubtless one of many which had gone off the feed and retired to a place of comfort and safety until the river became normal, or until conditions were once again favourable for feeding.

It is usually hopeless to fish for pike in a river that is rising or badly discoloured through the result of heavy rainfalls, though often they will take very well indeed during the two or three days just before the break in the weather. In this latter case it is reasonable to assume that the fish have an instinct which warns them of approaching high and dirty water—conditions which make it very difficult for them to obtain food. Knowing of these impending difficulties, the fish do their utmost to lay in a store sufficient to last them through a period of inactivity.

As the river rises and colours, the pike seek a sanctuary in places out of the current and there lie dormant—to all intents and purposes asleep, until hunger, or favourable feeding facilities prompt them to move. It would appear that these fish are capable of closing all respiratory organs and can lie upon the bottom while the mud and silt of the flooding water settles on and around them and they take on the appearance of their surroundings. A muddy log of wood is a good description for often the dorsal fin lies flat on the fish's back and its ventral fins are covered completely by mud. The eyes cannot be recognised as such and the whole body lies motionless.

While the flood waters swirl above and around him the pike lies hidden, safe in the knowledge that his cloak of mud is sufficient to protect and conceal him from his natural enemies. He has given up all thoughts of food, his eyes are sightless, glazed with mud, and his motive power is at rest. The instinct passed to him through many generations has told him that food will not be available, even though he hunts, and he is content to wait. Fishermen's baits may pass within a foot of him and even graze the very mud from his back, yet he remains immovable, dour and oblivious, while the angler, often unaware of the hopelessness of his task, goes hopefully to the next pool.

For some time I was continually being surprised to find big pike in water where during the previous autumn I felt positive I had left nothing of a relative size. I thought perhaps these were fish I had failed to discover in some deep hole here or there about the river, and that they had migrated to the place where I then found them. But I have since discovered that I was quite wrong in assuming this, for I have three times proved conclusively that a pike of 4 lb. can be over twice this weight in less than a year.

I can usually judge the weight of a pike to within a few ounces whether it is in the water or on the bank and twice during one autumn I missed fish which I took to be about 4 lb. I was using a harpoon and whether my luck or my eye was out I do not know, but I struck too high on both these fish, and, instead of getting them in the shoulder as I had intended, I gouged a deep furrow across their backs, just behind the head and left marks which I knew they would carry for the rest of their lives. And indeed they did carry the marks, for the following summer I got both these fish within a short distance of the place where I had missed them. Both weighed just over 8 lb.

The third example was a fish my assistant marked badly with a long-handled slashing hook. We were weed-cutting in September and as this pike, another 4-pounder, struck across the river, he hit it with the slasher. We saw the fish again the following day with a great gash showing whitely across its back, but it escaped into the deep water and we failed to find it again. That fish must have remained in the same locality until the following July when I caught it. Where the scar had healed was a deep furrow some $\frac{1}{2}$ in. wide, and it weighed $8\frac{1}{2}$ lb. I was surprised to see how perfectly the wounds of all these fish had healed over, for in each case in less than a year new scales had formed, though these were not uniform either in size or position. Judging from the manner in which the fish had thrived, it would seem that the wounds had caused little inconvenience.

I know it is a high rate of growth for a 4 lb. pike to double its weight in less than a year, yet here are three instances where undoubtedly it happened. But one must realise that these Upper Avon pike have the opportunity to live on trout, and in fact do so whenever they can get them. This probably accounts for the rapid way in which they put on weight, for though they may take four years to weigh 2 lb., the next four years may take them into double figures.

Some writers say a few pike in water do good rather than harm, as they eat all the sickly, ill-conditioned trout and live mostly on coarse fish. I think this is a happy way of tolerating the brutes, for though it may be quite true of the water of which they write, my experience on the Avon is quite different. For one thing, there are few sickly trout in the water even if pike wanted to take them, and, another, that I have repeatedly taken trout-filled pike from a pool where there were hundreds of suitably-sized coarse fish, such as roach and dace. I credit pike with having sufficient good sense to know good food from that which is second rate. Though he will take small grayling and other coarse fish, they are always in excellent condition, and I have never yet found a trout in a pike that was not a loss to the fishery. I wish there was such a thing as a good pike in the Upper Avon. I hate the sight of the brutes, for from my point of view the only good they ever do is occasionally to make a meal of one of their own kind.

On the other hand I see them from time to time at their crimes. One morning in May I stood on an old cattle bridge which spanned the river at a point where the water was about eighteen inches deep. Several two-year-old trout were feeding under the bridge and one in particular claimed my attention. The water was quite clear and, to get a better view, I gently lay down at full length with my head over the up-stream edge.

For ten minutes or so I watched the little trout with interest as it moved from side to side and darted here and there to secure the active nymphs that moved between the surface and the bed of the river. Occasionally a dun or a spinner floated along and was taken with apparent relish. The little trout was lying close to the bed of the river with its tail near a fair sized stone, and when taking surface food did so in much the same manner as a grayling. Long before the fly came within taking distance, I could see the eagerness with which the trout spotted it: the sudden check of the fins and the brightening of the eyes and the quick flick of the tail which propelled the body in perfect timing to the surface to meet the fly.

My attention was suddenly diverted as a movement some ten feet up-stream caught my eyes. I saw at a glance what it was, and a pike of about 3 lb. weight came cruising slowly down-stream. Now, I have seen pike in all of their many moods and I could see at a glance that this one was on the feed; a fish I would have liked to meet when I had a rod and bait. All the belly fins were moving gently, but the powerful tail and dorsal fins were held almost rigid, the dorsal curled slightly to one side. Its wicked eyes caught the light and glittered like jewels, and the whole fish seemed to be poised ready to strike in any direction. Slowly it came, on a course that would take it four to five feet to the left of the little trout I had been watching and which a glance showed me was still in position. That the latter failed to see the pike is beyond my comprehension; eyes that could see an olive nymph three feet away must surely have seen a thing many thousands of times larger, yet the little fish took not the slightest notice.

The pike, however, must have seen the trout, or at least have expected to find it there, long before it drew level, though it gave no indication that I recognised, nor did it alter its course. Then, when almost abreast of the trout, it happened. I was so close that I could have touched both hunter and quarry with a walking-stick, but I was too fascinated to move. All the fins and tail of the pike were brought into simultaneous action; a swirl of mud and gravel swept to one side and before I could think what was about to happen, the drama was enacted under my eyes. The trout did not move an inch before the gaping jaws snapped on its sides. I could see the head of the little fish on one side and the tail on the other as, without a pause, the pike turned up-stream and bore his victim off to a bunch of weeds some twenty feet above the bridge.

I could see a few scales glitter in the water as the cruel jaws moved the little trout into a position where it could be swallowed head first, and then the pike, now more than comfortably full, settled himself down. The eyes no longer glittered and all the fins lay flat upon the bed of the river; the fish looked as though in a state of stupor. I went home to get a wire and stick, and on my return found him in exactly the same place and an easy victim.

That was his last trout.

Now one might have thought that the trout would have rushed away to cover long before the pike got near, for there were plenty of hides nearby, but I have frequently seen trout show no fright on the appearance of a pike and indeed continue feeding. If it is apparent to humans that pike are on the feed, then surely it should also be apparent to trout.

I have given much thought to this subject, but still I cannot fathom it. I wonder whether it is possible that the pike exerts the same fascination on trout as a stoat does on a rabbit? When a trout sees a feeding pike close on him, does terror, in some circumstances, deprive him of the power of movement? From instinct, or from actual casualties amongst his kin, he must know that the pike is a killer. He has no doubt seen the coarse fish, such as roach and dace, scatter and flee when a pike strikes amongst a shoal of them. Frequently they leap far out of the water in the hope of escape—a welcome sight to pike fishermen and a common one to many anglers. Yet how rare it is to see a pike pursuing a trout; it is more usual, as in the case I have given, for them to make no attempt to escape.

I think it is extremely unlikely that pike are really natives of a running stream. Their habits are more suited to ponds, lakes, canals and slow-running water. I wonder if this is why the trout have none of the hereditary terror of pike that is shown by the roach and dace, for instance, who also, by their habits, seem to be out of their environment in a trout stream?

Sometimes I come across evidence that there is a chance to make war on the pike, as when I heard a distinct c-lop and at the same time some of the fronds of a big bunch of ranunculus near the side of the river lifted above the water. Something had moved there and, as I watched, I heard the sound repeated and saw a similar movement of the weeds farther up-stream. I then knew what was causing the sound and the movement, indeed I had expected to see such an occurrence in my wanderings that afternoon, and had my shot-gun with me. For it was the end of April and beneath those weed strands a female pike, with one or more attendants, was engaged in the process of spawning, of scattering far and wide the thousands of eggs she contained.

I quickly put cartridges in my gun and moved nearer. I knew from past experience that I would soon get an opportunity to shoot and that the fish would be close to each other, and so near the surface that a kill was certain. From the direction the fish were moving I could see they would enter an area where the weed growth was thin, and there I waited for their appearance. A hen pike of some 6 lb. came pushing through the weeds and to each side of her was a male of smaller size. The backs of all three broke the surface at once as they gave their convulsive spawning movement, and at that moment I fired. The shot killed all three as I had hoped it would.

It always pays to wander around the river or sidestreams at that time of year with a shot-gun, for I know of no other weapon so effective in destroying spawning pike. As many as half a dozen can be killed at a shot, for these fish have a habit of spawning near the surface of the water, and the males and female are so close together that they are often touching.

Though pike invariably spawn each year in waters such as the Upper Avon, the result from the pike's point of view is not always satisfactory, for though they may deposit their eggs and the eggs become fertile very often the temperature of the water is too low for successful hatching, and a season's egg-laying is destroyed. Pike eggs cannot stand highly oxygenated water; in this they differ greatly from fish such as trout and grayling. But there are seasons—seasons which unfortunately occur all too frequently from my point of view—when river conditions are favourable for the reproduction of pike, when eggs will hatch and the pickering thrive. At such times two or three pairs of pike can produce enough young to infest the whole river. Again a pair or so may find their way into a sidestream or adjacent brook, where temperature and general water conditions are at variance with the river.

I once assisted in netting a small pool where a pair of pike had been seen to spawn. The netting took place in September, when the stream leading from the pond had dried up. Actually the pool was isolated and escape for the pikerings impossible. From this pool—a pool with a diameter of about 15 ft.—we took a total of 323 little pike, some nearly 4 in. long. I give this instance just to show the number one pair of pike can produce. The parents of these were estimated to be about 3 lb. each, actually small size for spawners, and the hen was only capable of producing a comparatively small number of eggs. It is quite certain that many of the pickering we killed had eaten many of their brethren, for this is their habit, and it is also certain that, had this pool escaped attention, most of the fish we destroyed would have dropped down-stream as soon as sufficient rain, or spring water, linked the pool once more to the river.

Many and varied are the theories and facts that are put forward to account for the activity, or otherwise, of pike, for it is well known to fishermen that there are days when they appear to be ravenous and will take almost anything that has the slightest resemblance to the prey they are accustomed to seeing.

My own views are that pike are greatly influenced by changes in the temperature of the water. I have many times had proof that in winter or spring pike will feed in a river when a mild spell follows severe frost and snow. In summer and autumn, it is just the reverse; a fall of two or three degrees, caused by wind or rain, or both, will start the pike moving about the river in search of food.

My observations are not based so much on fishing as on the habits of the fish, though I have caught many hundreds of pike with a rod and can well remember the best days and the conditions which made them possible. In summer and autumn I hunt pike with wire and harpoon, for the clear water of the chalk stream gives one the advantage of being able to see much of the river bed, even in the deep pools. There have been days during a hot summer when, from the banks, I have hunted a reach of water from end to end; when the blazing sun and cloudless sky has made it possible for me to see every stone in the river bed, and yet failed to see a single pike. Yet I knew pike were there. I have waded that same stretch shortly afterwards, walked through the weed-beds, poked under banks and tree-roots and every place where a pike could lie concealed. And there they have been, lying absolutely inactive until I have trodden on them, or poked them with the end of my probe. No bait would have moved a single one. Yet, on the other hand, I have been along reaches on a day when a hot spell has broken up—when rain and wind have been lashing the surface into a turmoil and when the ripple on the water has made it difficult to see beneath it. There, as if knowing they were quite safe from my wire or harpoon, I could catch glimpses of the pike as they lay out in the open with glittering eye and waving fins, waiting for some unsuspecting prey.

There have also been days during the winter when I have fished in the most severe weather—when my rod rings have frozen solid, and the line has become a block of ice on the reel. Not once, but a hundred times, until I learned at last the futility of trying to catch pike in such conditions. I should have done better had I used a trident and gone prodding into the rush-beds at the edges of the water, for that is where the pike were lying and, by chance, one might have fallen a victim.

Then there have been days when a warm spell has followed days of frost and snow. Pike have sensed the rising temperature of the water. As with the change during summer they know that their prey have also been awakened and are likely to be obtained. So shaking the mud from off their scales they sally forth into the stream to lie at some vantage point and wait for what Nature has to offer. And at such times, should a bait pass within sight it is taken without thought of the consequences, for the pike are on the fin—just waiting, ready and eager. The pike fishermen should try to take advantage of these feeding times and I have often wished I could be in a dozen places at once.

Pike rely on their sense of sight to enable them to obtain a meal, do they not? Would they die if they became blind? I was asked these questions one day and I answered that a pike would most certainly die if it became blind, but not immediately after the blindness occurred. For blindness is the ultimate natural ending of the lives of most freshwater fish, and probably also of those in the sea. Pike, however, are likely to die more quickly than many other fish as they rely entirely on their sense of sight to find food. I have killed pike which were totally blind—poor, emaciated creatures scarcely more than head and fins—and I have killed many others that had lost the sight of one eye, yet were still in good condition.

Only a short while ago, armed with harpoons, my assistant and I were hunting a reach for pike. The weeds had recently been cut and it was quite easy to see the bottom even where the water was deep. A pike of about 8 lb. struck out from the bank and lay in full view in the middle. Now pike in a running stream usually lie with their heads facing up-stream and their bodies perfectly parallel with the current of the river, and I was rather surprised to see that this fish, even though it was in the current, was lying at an angle with his head facing towards my bank. My assistant slid his harpoon into the water and tried to get him from his side of the river but the fish was scared and bolted up-stream. It was quite easy to follow him along, and when he stopped he was within easy reach from my bank, but again he was lying at an angle and, as before, with his head facing towards me.

I approached cautiously to within striking distance and

got him with my harpoon. As soon as he was on the bank
I could see the reason for his strange behaviour. The eye
which had been nearest to me in the water was totally blind,
though the other appeared to be in perfect condition.

Some people say that fish have no power of thought.
They may possibly be right, but it would appear that this
pike knew of his handicap and that he knew perfectly well
that he was being hunted. For without question the pike
was doing his best to lie in a position where his one eye
could command a view of both banks and the whole
area of water between the banks in front of him. The fish
was in excellent condition and had recently taken a 12 in.
trout, so it would seem that he experienced little difficulty
in obtaining a meal even though he had but one eye. My
assistant said I took a very unfair advantage by harpooning

the pike on his blind side, and put his own failure down to the fact that in all probability the fish had developed a very keen sight in the one sound eye which was looking towards his side of the river, both down- and up-stream.

Though perhaps I did unknowingly take an unfair advantage, I have no scruples where pike are concerned. And after all, I did end the fish's life while it was in good condition, for undoubtedly the sight of the other eye would have been affected in the course of time, and it would have died a lingering death through starvation.

13   WEED AND WATER

FEW things can be more pleasant on a hot day than to sit at the riverside beneath the shade of some overhanging tree listening to the water rippling on its journey to the sea, or to peer into the soothing depths of a placid pool. Indeed, I would be the last to want to see all bushes and trees cut away from river banks, but the value of direct sunlight on to a water is a point that needs careful thought. Listening to the music of water from a peaceful, comfortable spot is all very well, but often this same water is expected to breed stocks of food for trout, and to produce the many insects which make possible the sport of fly fishing.

Many people like to imagine that overhanging trees and bushes are themselves providers of trout food and that a continual supply of succulent creatures are forever tumbling from the branches to sustain the eager fish living in these shaded positions. But there is no fact to support such theories, for few are the creatures which fall into the water, and their food value is negligible when compared to the multitude of life which would be present, had not the water been shaded throughout the months of spring and summer.

Trout will live in these darkened areas, in fact, the larger fish often prefer the seclusion provided by an overhanging branch or bush. But their reason for living there is one of self-preservation, rather than because of food supply. Such positions are little more than daylight holts, a home into which they can retire and feel comparatively safe in time of need. A certain amount of food is brought down by the current, but to obtain the abundance and variety that is necessary for their existence, they have to range up-stream or down to water that is open to the sunlight.

The life of trout, and indeed the whole nature of a river depends on the ability of the water to create and sustain the first minute life, the protoplasm, the creatures that are invisible to the naked human eye. Though these microscopic animals must have water in which to live, water cannot alone bring them forth. A combination of *sun and water* is necessary, for with one or the other alone they will cease to exist. Many millions of creatures must die to produce a trout of $1\frac{1}{2}$ lb. in weight, because his existence depends entirely on the life of smaller creatures, and in turn the trout's food must also have something on which to feed. And so it goes on in one vast cycle all down the scale to the minute organisms which must have sunlit water to give life and to foster them.

So when thinking of opening up water to the rays of the sun, one must not just think in terms of trout, or even of immature fly life. There could be no trout, or insects, but for the protoplasm, and if water and sun are allowed to combine to create this, then other creatures will live and thrive. It is the vegetation which, screening the water from the morning sun, is the most harmful, for Nature expands from dawn to midday and rests during afternoon and evening. A water cannot be productive if it is robbed of its assistant, and, though trees can add to the beauty of a riverside, and act as wind-breaks in time of storm, they can also form a barrier and a deterrent to the creatures of the stream. There is an old saying that if you look after the pennies, the pounds look after themselves. The same thing applies to the river. It pays to look after the little things. The big ones are sure to follow.

Unfortunately there are times when the good growths of vegetation in a river are destroyed. They may be choked with flannel weed, killed by pollution, lack of aeration, or other causes. Where these growths are considered to be assets to the fishing, it becomes necessary to give Nature a helping hand to re-establish beds of the better classes of weeds.

Transplanting weeds is not a very difficult operation, but some thought on the subject is needed before one makes an attempt to do it. The first, and most important, thing is to have knowledge of the kind of vegetation the water is capable of producing. It is much better to try to increase the native stock rather than to introduce classes of plants from other waters. One can be guided by Nature, for she has already decided which vegetation is most suited for a particular water, and at some time or another such vegetation has flourished.

In the south country where water is alkaline, ranunculus and celery are considered to be of most advantage in a trout stream. Both are quite easy to transplant, and when the need arises, I use them, in preference to all other types, to provide a quick-growing shelter and feeding ground for the river creatures. A second point to be considered is the class of water and the condition of the river bed. Ranunculus and celery will both thrive in well-aerated, shallow reaches, and in water that is of fair pace. Indeed certain aeration and pace of water is necessary for their well-being. They will flourish and become abundant provided there is sufficient fertility in the form of humus and soil in the river bed into which the roots can thrust themselves to feed the plants and get a grip to maintain themselves. I usually make an examination of the river bed at the points where I want to replant. I test the bed with a garden fork to see that there is a certain solidity. I find that if it is possible for me to dig up a spit and lift it to the surface without it sifting through the tines of the fork there is sufficient soil in that part to give weed growths a chance to thrive after I have planted them.

Where the stream is strong, it is necessary to protect the roots of the newly planted weed so that currents cannot wash away the loosened packing around them. March and April are the months when weed growths throw out new roots and start reaching their fronds towards the surface of the water, and this is a good time in which to replant. The plan I have adopted, and which I have found to be successful at various times, is to get a number of sandbags and treat all replants individually.

First of all, after finding beds where the young growth is healthy, I put a mixture of clay from the river bank and silt from the river bed into the bottom of a sandbag, sufficient to make it about one-quarter full, then with a garden fork I dig up a section of the river bed in which there is a plentiful supply of weed-roots. These are put into the bag on the top of the initial mixture and pressed well down into it. Then add enough of the first filling to well cover all roots, and press firmly. Any long strands of weed should be allowed to hang out of the mouth of the bag.

The replant is then ready to be placed where needed. A hole sufficiently deep to take bag and contents should then be made where the formation of the river bed is suitable. The whole bag can then be inserted with the mouth part facing down-stream. Fill in and pack tightly around the bag, then tread all down until on a level with the river bed. To complete the operation cut away the surplus bagging at the mouth without damaging any trailing fronds, then split the surface of the bag lengthwise to expose the contents to the light. The two ears of loose bagging can then be folded back to either side, and weighted down with a few small stones.

This method can be adopted with any type of weed and in all classes of water. Where water is fast the bagging successfully prevents any scour direct on to the roots which may wash away soil and expose them. In time, the sacking will rot away and provide a humus on which the plants can feed.

After the Mayfly is over I usually carry out the first

general weed-cutting of the season. For a week or two small flies are scarce. Fishing in the daytime is not likely to be good, and it is quite time the job was done.

If one could consider river vegetation solely from the point of view that it is valuable as providing feeding places and refuges for insect life and trout, the job of weed-cutting would be a simple one. But there are many other factors to be thought of, for a hiding place for trout can also be a place of refuge for pike; a feeding site for insects can be a harbour for mud. Prolific weed growth is a serious obstruction to the passage of water, and can be a nuisance for dry-fly fishing. Careful and limited weed-cutting is, however, a question that should be thought over methodically with a view to making the best of everything, for without doubt the vegetation of a river can play a big part in the success of a fishery.

No matter how it is done, weed-cutting is hard work, for at present there is no machine capable of dealing with classes of water where the river is harnessed, and where there are alternating deeps and shallows. Everything has to be done by hand, either with chain scythes (a series of scythe blades linked together with a rope at each end for pulling); by pole scythe, a scythe fitted to the end of a long pole; or by a hand scythe—mowing tackle as most of the old farm-hands call it.

Both chain and hand scythes are necessary. To use a hand scythe one must be in the water and it is hopeless to try to cut weeds where the water is more than three feet deep. This class of water has to be tackled from the bank and here the chain scythes are useful. Where fishing is possible I try to leave about one-fifth of the vegetation growing in the deep water and about one-third in the shallows. But where the nature of the river is such that fishing is impracticable, then I just mow sufficient to allow passage of water and drifting weeds.

I much prefer to mow by hand, for then it is possible to select the most suitable growths and to leave them untouched. These weeds can be arranged so that they alter the course of the stream and throw the current from side to side, while in general effect they create an aerating action which is so beneficial to insect life in all stages, and to trout. It is also possible to leave a short growth on the bed of the stream and, in general, create an appearance that is both useful and pleasing to the eye.

With the chain scythes it is all or none, the linked blades skim along the bed of the river and cut everything before them, for to be successful in their use one must necessarily work up-stream, so that the blades can slide beneath the fronds which trail with the current. Deep waters, where I employ chain scythes, are usually the homes of big trout, and often of big pike. Here the weeds can be cut cleanly with advantage, so that when an angler hooks a big fish there is a good chance of being able to land it. And it may be possible to spot a big pike whose presence is far from desirable. In the cleared water one may get him. But when I say cut deep waters clean, I mean only the bed. I do not mean that one should start at the bottom of a three hundred yard reach and skin out every weed, unless one desires to net the water, for this is a great mistake. In such a reach I should try to leave five or six bars of weed across the river, and to cut cleanly between them.

Well trimmed river vegetation adds to the charm of a river, for who can look at the myriads of buttercups on a bed of crowfoot, or the bright green of starwort, or celery, Nature's flower-beds, without being struck with their beauty? And when one knows these lovely water weeds are the refuges and feeding places of insects which are equally lovely when they appear as flies about the valley and of the bright-spotted trout that may give sport, then, though a day of weed-cutting can be hard work, it has its satisfactions.

Though masses of cut weed continually floating down a river can be a curse to anyone fly fishing those same masses have a value that canot be overlooked. Run a fine-meshed net through a buch of ranunculus and many thousands of insects can be collected in a very short time. Much of this

life is in and on the weeds, well above the river bed, and in
cutting adrift vegetation one also cuts adrift the living
matter in it. Most of this insect life is of fly families which
lay their eggs in aerated water, namely on or near shallows
where the best classes of river weeds live and thrive. Here
the eggs hatch and the very tiny larvæ find food of a kind
they require, and many of them develop into nymphs. So
one can look upon these shallow waters as the breeding
places for many of the insects which make dry-fly fishing
possible. Though well advanced larvæ and nymphs will
thrive in deep, sluggish water, there is little food of a nature
suitable for freshly hatched larvæ, a fact that is undoubtedly
known by the parent flies, which rarely deposit eggs in such
water.

So on these shallows, amongst the weed growths, are
insects in all stages, some nearing maturity, and in quantities
out of all proportion to other parts of the fishery where
hatches of fly could provide sport. Such shallows may extend
for several hundred yards, and down-stream there might be,
by way of contrast, a similar length of sluggish water where
few naturally bred flies are to be found. Weed-cutting acts
as a stocking method for this latter class of water, because
when weeds are cut and allowed to float down-stream, they
carry with them insects which are sufficiently advanced to
find food, and to develop in the deeper water. In this way,
a reach, which may otherwise remain barren if it had to
depend on egg-laying, is stocked with insects which will
eventually hatch there and, in so doing, will provide an
attraction to bring the trout to the surface to feed on them.
So, though floating weeds may spoil a day's fishing, they can
create facilities whereby sport on future days may be
intensified. For as the cut weed drifts with the current, this
fact is discovered by the inhabitants and, having no desire
to be carried away from their homes, those that can will
quickly get clear. All the way down-stream these larvæ
and nymphs are struggling to get free, and leave the drifting
masses as soon as they can. If growing vegetation has been
left at intervals in deep and shallow waters, the insects

soon find a new home, and there they live until they are ready to hatch.

In some waters, the cut weeds are immediately cleared so that there is no interference with the fishing. The vegetation is either by-passed into some disused sidestream, and there left in heaps, or pulled bodily from the water to rot on the river banks. I think this is a great mistake where fly fishing is practised, for in either case, the very life that provides the sport is being needlessly destroyed. If it is essential that weeds must be removed, an effort should first be made to pass them through some rapid water, where the masses could be broken up, and a chance given to the insect life to escape.

When weed-cutting is done I try to pay visits to some of my keeper friends who look after fisheries on other rivers. Keepers, river keepers especially, often work single-handed throughout most of the year, being assisted only by casual labour when extra man-power is necessary to do some of the heavier work such as weed-cutting, bank repairs or netting, and for this purpose employ local men who are quite willing to do as they are told. So many of them get no opportunity of seeing the methods of others who are engaged in a similar occupation. Generally speaking, a river keeper lives a life apart from his kind, with only the wild life around him for company, and judging by my own experience, he welcomes a chance to see other rivers and to talk over the many interesting things that are part of his life and work. Views of other keepers are listened to with eagerness in the hope that some benefit may arise that will be helpful in the future, or that an explanation of a natural event will clear up something that has been a puzzle for many a year.

Though we gain knowledge year by year, the advance in the general upkeep and management of fisheries has been very slow, and, looking at it from a sensible point of view, this is easy to understand. The men who do the work, and by their efforts or lack of them can make or mar a fishery, have, in most cases, to follow their own inclinations. River keeping is not just a matter of walking along the banks to prevent any unauthorised fishing. It involves a multitude of different jobs and the river keeper must be a jack-of-all-trades, and master of a good number of them. There can be no hard and fast rules for his guidance, for varying waters need different management and his experience has to be gained through trial and error.

As a rule, it is the employers who do the visiting, see the work of other keepers and question them concerning it. Occasionally they get useful information and can pass it on to their own employees, but many keepers are too modest, or shy, to talk confidentially to any other than one of their own kind, and even if they did talk there is such a thing as a keeper's language, not bad language, but a manner of speech that can only be interpreted by a kindred spirit. In most occupations, whether purposefully or indirectly, men help one another, and advance in their particular line is continually being accomplished. Take, for instance, factories, or buildings, or agriculture, in fact, any work where a number of men are continually together. The success made by one man is quickly noticed by another, and often a thoroughly skilled man is present to correct any mistakes and to give help in time of need.

Keepers must necessarily have a knowledge of Nature, and throughout their life they have endless opportunities to study the wild creatures of the river and riverside. Many have done so, many have found out little things that are of great value and which could have filled a big gap in our river knowledge. Those men would have been pleased to talk to someone equally interested if they had been given the chance, but most of them grew old in their own little spheres and have taken their observations with them to the grave. I am encouraged to visit other rivers and to talk with other keepers; I see other methods of work that are helpful, and if I can in turn give advice on any subject I do so. For only by helping one another can we hope to learn of Nature's many secrets and by doing so to break fresh ground instead of continually going where others have trodden before.

Many people think that to catch fish all they have to do is to get a net and drag it about the water. How sadly these same people are disillusioned when they make the attempt. The successful netting of any water is a highly skilled job, especially in rivers, but wherever it is attempted, certain preparations and forethought are necessary, and the work must be carried out in a methodical manner.

Some months ago I watched the netting of a certain water where there were hundreds of coarse fish. This was being done by members of some angling clubs. It was the intention that all fish caught were to be transferred alive to club waters, and transport vehicles with tanks were standing by to receive the fish as they were captured. Apparently some good hauls had been taken from this place at some time earlier, and as I studied the water I could see that in it was a shoal of roach, rudd and perch, which must have amounted to many thousands. The thought crossed my mind that many of these fish had escaped the previous operations. Soon I knew why.

It was an easy place to net. At one end was a deep pool, and I knew the fish could be worked quietly into this pool, and then encircled with the net. In fact, it would have been a simple matter to trap the whole shoal and land them, for the bed was even, and quite clear of vegetation and snags. I watched with interest as the men started the job, and at once I could see that two or three of them had a good idea how to catch the fish, and that they were doing what I should have done. The net was a good one of about one hundred yards long and some fifteen feet in depth. It was well leaded and corked, but it was of small mesh and heavy. Great strength was needed to move it through the water. However, there was no lack of man-power and all were eager to help.

But this is where most of the trouble arose. The men were too eager, and all knew better than his neighbour how to do the job properly. Advice from one and another was continually being shouted across the water, and soon there was confusion. But it seemed to me that the fish were quite willing to have a game. Probably there were some old hands at escaping nets amongst them, and they knew they were quite safe. Without any trouble, the whole shoal moved into the deep water of the pool, and were encircled with netting.

The hard work of the operation was done. The fish were successfully trapped and the critical time had come when the shoal had to be brought to the bank. This was the time when the fish would fight for their liberty and take advantage of any escape route that was offered. The moment had come when it was necessary that one man should direct operations, with the whole gang working in union under his instructions. But here everyone had his own ideas, and the one or two who understood the work were ignored.

At each end of the net, a dozen or so men, all excited by the sight of the big shoal milling about in the half circle of netting, hauled on lead line and corks with the one thought of heaving the fish to the bank. Many others were getting in the way with buckets and baths to receive the catch. Everyone was most concerned when a fish or two slipped over the cork line to freedom, but none of them gave a thought to what was happening to the lead line at the bottom, or if they did, nothing was done about it. Soon it rolled up so much, that with every pull from the bank the whole net lifted from the bed and, in consequence, a gap appeared beneath the leads through which a shark could easily have passed. The trapped shoal were not long in taking advantage of this avenue for escape.

It was quite obvious to me that soon the whole mass of fish would be gone, but nothing could stop these excited anglers. As they pulled the net rolled up until it resembled a great rope, and as the area of netting decreased in the water so did the numbers of fish. When finally they got the net to the bank, there were but five small, unlucky fish trapped within its folds.

Here two hours of hard work was wasted in less than ten minutes, and a haul of fish worthy of transferring to any angling water had been lost through too much haste in landing the net. Fish never get unduly alarmed until they find themselves trapped in a small area of shallow water, and before this happens the netting should be worked so that it forms into a kind of horseshoe shape—a formation with cork line riding the water at a point almost directly above where the lead line is resting on the bottom. It is important that the lead line is clear, so that, as the lines are pulled, the loose netting will belly out behind and so form a bag in which the fish, though trapped, have water in which to swim. Then with gradual movement the whole net can be worked in to land, and bring with it the catch.

There are leaders of fish as well as of men, but with fish it is always one of exceptional size who takes the rôle of commander, so whether brain and brawn go together in fish life remains to be proved. This fact of leadership was brought home to me once when we were netting a reach for coarse fish and pike. Weeds had been cut and, though it was then early November, it was still possible to see the river bed in many places, and to see the fish as they moved about over it.

Roach and dace seem to like each other's company. They can often be found together in somewhat equal numbers. Such a shoal was in this particular reach and, indeed, there were about 500 fish of all sizes up to $1\frac{1}{2}$ lb. But amongst them was a roach in a class of his own, and, viewed at times, he appeared to be twice as big as any of the others. Many times during the summer I had seen these fish and whenever I disturbed them, the big roach would take the lead and the others would follow wherever he went, whether up-stream or down. He was there with his following when we were ready to start netting, and I was anxious to get him to see how much he weighed.

We were using a stop-net and a drag-net and, following our usual custom, we set the stop-net at the lower part and took the other to the top of the pool to drag it down-stream. It meant a drag of about two hundred yards and when we started I knew that all the shoal were between the two nets. All went well until we had got to within fifty yards of the stop-net, and then the lead line fouled a sunken branch at one side and up it came almost to the surface. It was a critical time to have anything like this happen, for I knew the fish had already discovered that their down-stream progress was screened. Though we quickly cleared the snag, I was afraid an avenue of escape had been offered. The moment we landed the stop-net my suspicion was confirmed. At least half of the shoal had escaped and with them the big roach.

I knew he must have gone up-river so, quickly putting the stop-net in position again, I decided to make another drag. On the way up-stream along the bank with the drag-net I tried to locate the big fish and his companions, but the water was still muddied from our operation and I could not see very well. Now, just up-stream from where we started the first drag, was a shallow reach of about one hundred yards. To pass over this I knew the fish would have great difficulty and that in fact they would have to swim with their backs out of water and I thought it extremely unlikely that they would do so. So feeling confident that the big roach and the others were still in the deep water we netted it a second time. It was not quite a waste of time as we got a pike of 2 lb., but, though we had a good haul of trout, there was no sign of roach or dace.

I still felt positive the shoal had not escaped the nets this time, and was equally sure they had not gone down the river. So leaving the nets on the bank with my assistants I went up to the shallow. Almost at once I saw a wave approaching from where the water was a little deeper up-stream, a distance of nearly a quarter mile from where the net had been snagged. It was the big roach, and strung out behind him were the remainder of his band. Without a pause, they all scurried down over the shallow with their backs out of water, splashing and struggling in their efforts to get to safety, as they thought, in the deeper reach that had been their home.

I shouted to my assistants to put in the stop-net and bring the drag-net to me while I kept watch to see the fish did not pass up-river again. For the third time we dragged. Everything went as it should and I was certain when we pulled in the stop-net, that we had the shoal trapped. We had, and, as the net drew nearer to the bank, one of my helpers spotted the big roach. With an idea of making sure of the fish he dropped the cork line he was holding, and grabbed it. For a moment he held the old roach high enough for us all to see, then it slithered from his hands and, with a plunge, went over the corks to freedom. I would not try for him again as I felt he deserved his liberty.

It was exactly twelve months later when we got him. He was again a leader, but this time he looked enormous amongst his following, as these were fish of about 6 in. long. They had escaped through the net meshes the previous year when they were fry. I made an individual effort to get him as I had come to the conclusion that the old monster knew as much about nets as I did. He was a grand specimen and weighed just short of 3 lb. Without their leader the small fish became easy victims and we successfully netted the whole shoal.

Many of the mysteries, which from time to time have appeared in print about fish and fishing, have very simple solutions when you are able to look into them from a natural point of view. During the winter of 1951, people who were observant and who studied the rising springs, the flooding and general high water about the countryside, no doubt found a solution to a mystery that has puzzled many fisher-men, and others. I write of the fact that fish of various species have been found in ponds that are long distances from the nearest waterway—ponds which have been known to have been perfectly dry at some time in the past, and where there is certain knowledge that no fish of any kind have been introduced by human agency.

Some say, by way of explanation, that the fish may have been carried to such places (in egg form, as alevins, or fry) by herons, wild-fowl, otters, or frogs. Well, this is a possible solution; fish could be transported in this manner, but there is another, a natural and more practicable, way in which fish can get into the pools. From time to time we have periods of drought and then of high water. During the former, these ponds are cut off entirely from other water.

It is usually during dry seasons that fish are found to be occupying them. The absence of any overflow, any connection with other waters, or any visible sign that there ever has been a link with stream or river, gives rise to the belief that to get into the pond the eggs or fish would have to be carried, or that the fish have to travel over dry land.

But let us look at these ponds during a winter such as we experienced in 1951. Springs rose to a very high level and heavy rains and thaws of snow came in addition. Ponds and pools everywhere filled and overflowed until all waterways were linked one with another. In fact, a stream of water flowed from pond to river, a stream carrying sufficient water to allow fish of good size to travel in it, should they have that desire.

During winter fish will migrate, especially the small ones. A shoal of dace or roach, an odd pike, trout, minnows, or indeed any species of freshwater fish, at times do have the urge to explore virgin waters. The crystal-clear spring water has an attraction for them, they get into the stream which comes from the pond, forge away up the watershed eventually to enter the deep water of the pool. There, finding conditions to their liking in that there is a good supply of food, they settle down to enjoy themselves.

There they stay, quite happy in the clear, pure water, but as winter turns to spring there is a gradual change. The flow from the springs gradually lessens. Soon there is no overflow to link the pond with the stream or river. Soon there is no sign that such a thing has happened at all. There is just a pond which to all visual appearances is isolated and cut off completely from all other waterways.

Years may pass before the pool once again fills to overflowing, in the meantime there may be drought. The pond water sinks lower and lower and the fish are discovered. I have seen many of these ponds fill and overflow, I have also seen them dry out, and remain dry for years. I have seen fish enter them and I have seen them stranded when the water has all drained away.

14   SUSPECTED PARTIES

From time to time throughout the year the changes in the season and in the routine of my duties gives me a chance to make a note of the effect on my feelings. For instance, to me November is a sad time, the one month of the year when at times I feel lonely on the river, and no doubt other river keepers feel the same. No longer have we the glory of autumn to occupy our thoughts and keep them from the grim picture of winter, for each day brings something to remind us of the cold grey months that must pass before we once again see nature awaken. As I tramp along the banks my feet rustle through heaps of dried-up leaves— leaves which but a short while ago were clinging, in all their beauty of autumn tints, to trees which by then are lifeless and silent, with gaunt branches reaching towards the sky. All the summer migrants have gone to warmer lands, and our own winter birds are far too busy hunting for food to think of brightening the day with song. Dead rushes and willow herb, killed by the early frosts, droop along the edges of the water, while here and there a tangled mass of figwort is supported by the stiffening stem of a teasel.

For the most part the cold, dark-looking surface of the river remains unbroken by rising fish, but here and there a grayling may rise to take one of the few duns that are hatching. No longer can I see the bright spotted trout lying on the gravel behind a weed-bed waiting for food, for now the water is discoloured with recent rains and beds of ranunculus have died away to short growths, which are pressed tightly to the river bed as the increased water supply travels swiftly towards the sea. But there is still work to be done, and work is a good cure for loneliness.

I always find it far easier to kill pests at a time when they are doing the most damage. I then feel a resentment towards them and for the time being can forget any interest the creatures have provided for me in other seasons. A mouse might make a pretty picture when it sits up and washes itself on one of the kitchen cupboards and be left unmolested, but when that same mouse starts to nibble the cheese ration his death warrant is sealed. Grey squirrels may be very interesting sitting on a tree stump or swinging from branch to branch in the autumn, but when an absence of song birds is noticed in the spring one hates the sight of the animals and goes all out to kill them. The same can be said of many of the creatures we have come to regard as pests and amongst them is the dabchick, or little grebe.

Many an interesting time have I spent watching a dabchick and her young, and many a chuckle have I had when seeing a fisherman casting his fly carefully over a dabchick "rise," especially when I could see the cause peeping from behind cover, where it has surfaced a second time a few yards away, with its beady black eyes and white-gilled mouth, appearing as though the bird is grinning, and thoroughly enjoying the joke. Indeed, I have a certain regard for the little diver, pest though he is.

In waters where fishermen have to rely on natural-bred trout for sport, the small ones in the river are just as important as those that are larger. The sport for future years has to depend on these small fish attaining a good size. Even though the whole life of a trout is precarious, the most critical time in their existence is, I think, the first year of their life. This is a period when they may become the victim of one of numerous enemies.

It is during this early life of trout that the dabchick can be destructive for, though I have killed dabchicks that have taken trout up to yearling size, I have had no evidence that they attack anything larger. Dabchicks, as the name suggests, are divers and take most of their food from beneath the surface of the water. They are fond of trout eggs and alevins, and will hunt systematically through the spawning redds to find them. They will also search the edges of river or stream, and take the little trout fry as they lie unsuspecting in their feeding places. Trout up to 4 in. long fall victim as the little diver searches river bed and vegetation, and so in the interests of fishing the number of these birds must be kept in check.

Usually, by the middle of February, the sight of dab-chicks about the spawning grounds of trout has incited me to take action, and often February is the best month of the year in which these pests can be destroyed. Generally, the weather is frosty and cold, with mornings when the grey mists rise from the river like steam from a boiling pot. Dabchicks have no great liking for diving in ice-cold water at early morning and a walk along the bank will find them on the surface of the river with feathers puffed out like a broody fowl. Cloaked in the mist, approach to gun-shot range is simple, and they become an easy prey.

One might perhaps come upon some that are feeding— diving and surfacing at regular intervals. It then pays to wait until the little grebe has dived, then run forward a few paces and crouch, wait until the bird has surfaced and sub-merged again, then repeat the move forward until such time as an easy shot can be obtained. Sometimes, however, two or more may be feeding together and one often surfaces as another dives, as though taking it in turns to watch for danger. Approach then becomes difficult but if one is patient, there comes a time when the diving bird finds some choice tit-bit, and his mate, becoming impatient, dives once more to the river bed regardless of his turn as sentry.

At moulting time my fowl run becomes a mass of feathers and some of the fowls present a queer sight with new feathers sprouting on all parts of their bodies, legs and wings. As I look at them in this condition it suddenly struck me that of all the wild birds I see it is only the moorhen which undergoes a complete annual moult. When one gives some thought to this subject it is quite obvious that moulting would place other birds at a decided disadvantage, for, like the moorhens and domestic fowl, they would be unable to make use of their wings to escape from predatory creatures.

I had often wondered why moorhens, being mostly vegetarian, should have developed the habit of diving beneath the water to hide and to swim, for, unlike the little grebe, or dabchick, they do not feed much on aquatic creatures and seldom, if ever, dive to the river bed for food. Then it occurred to me that this habit is possibly one for protection only and was developed mostly as a means of escape when the time for moulting is at hand. There is a period of at least three weeks in late July and early August when the primary and secondary wing feathers are in the process of growing, and during this time the moorhen cannot fly. The only safe refuge they have is beneath the water, and many times I have seen them dive and cling tightly with their feet to vegetation on the river bed, while tiny bubbles of air escape at intervals from their beaks to rise to the surface like little silver balls. Sometimes they are aware of my presence. They can see through the clear water and know I am watching them. One day, out of curiosity, I timed one just to see how long it could remain submerged. I knew it could see me, and though movements occurred from time to time which sent up a shower of bubbles from the feathers on its back, it was exactly six minutes before the head of the moorhen broke through the surface of the water. The bird was completely exhausted and for some moments it gasped for air before diving again.

They do, of course, dive at other times of the year and also are good at swimming beneath the water. With rapidly kicking legs and with wings beating the water as though in slow motion flight, they can make good speed. I have known many to swim a distance of thirty yards or more before surfacing and all the while a chain of tiny bubbles has marked their progress. With this chain of bubbles one can often determine whether it is a moorhen or a dabchick passing beneath the water, even though you have not seen it dive. Dabchicks move much faster but seldom send a continuous chain of bubbles to the surface.

Writing of dabchicks reminds me of a nest I watched one July. How quick these little birds are in covering up their clutch of eggs when they have been disturbed whilst sitting. It seems incredible that the eggs can hatch successfully amongst the mass of wet weeds of which the nest is constructed, and a casual observer can be forgiven for thinking the eggs must quickly get cold, and remain cold, for often a dabchick is away from her nest for an hour or more after being disturbed. At one time I thought the same, and that the mother bird covered her eggs to prevent them from being seen, but now I think it far more likely that it is done to keep them warm. The nest of a dabchick is constructed almost entirely of freshly gathered river weeds. These are all packed together into a roughly shaped ball with a little hollow for the eggs just above the water level, more than three-quarters of the nest being submerged. Tightly packed river weeds are similar to tightly packed green grass. As with the grass, the centre part becomes warm as the weeds begin to rot and, as the farmers say about a wet rick, it begins to heat.

When a dabchick covers her eggs it is like shutting the door of an oven. She is sealing in the warmth which her body and the weeds have created. Though the nest of the little grebe may look nothing but a sodden mass of cold, wet weeds, this is but an illusion. Instead, here is a cleverly thought out plan of incubation and one that is entirely successful.

To me a river without a moorhen is like a farm-yard without a sparrow. Though these little wildfowl can at times be a nuisance during the fishing season, I should be very sad to miss the sight which is now so familiar, for moorhens are plentiful on the Upper Avon and many times they have provided me with enjoyment and interest. One June I was very amused while watching an episode of family life with a pair of moorhens. The hen was sitting. Her clutch of eggs was shortly due to hatch and I think she wanted to extend her nest to accommodate the chicks. It is the duty of the male to find and bring this extra material to the nest while the hen remains sitting, and apparently this cock was anxious to please. Repeatedly he made journeys to and from a rush-bed on the far side of the river, a distance of about fifty feet, and carried pieces of dead rush to his mate.

There was plenty of similar material within a few feet of the nest, yet, for quite ten minutes, the cock went across and returned from the other side of the river. Perhaps he tired of doing this. If so, I cannot say, but, moving cautiously up-stream, some ten feet from the nest, he took a piece of rush and went back to the hen. She must have been watching his movements. For a while there was some altercation between them, and he repeatedly lifted up the fragment in his beak for her to take, as she had the others. But suddenly she snatched it from him and, as I thought, deliberately dropped it into the water.

Very slowly he swam across to the other bank and returned with a peace offering, but shortly afterwards he went up-stream again, and returned with another fragment from the nest side of the river. This time there was a show of distinct annoyance. The hen rose up on the nest and squawked at him and, though he lifted his beak and offered her the dried rush, she refused to take it, and turned her back to him. He then moved away and, still carrying the bit of material, went across the stream to the rush-bed on that side. It was fascinating to watch and, as he pushed his way into the rushes on that side, I wondered if he was going to sulk. But no, he paused a few moments, with the bit of rush still in his beak, then, turning round, out he came and sailed across the river. Without hesitation the female took the fragment and began weaving it into the nest.

He turned away. Perhaps at that moment the light had changed, or I had a view from a different angle, but I am positive there was a twinkle in the eye I could see. Once more he sailed across the stream, but it seemed that he then had made up his mind to obey orders, and to get all nesting material from the distant edge of the river. Though I watched for more than an hour, while he worked continuously, I saw no further attempts at deception.

I have written this little incident because, apart from the comedy of deceit, it produced ample evidence that moorhens have the power to give thought to the future. This sitting bird, in giving orders to her mate, was planning for the time when her chicks would need fresh accommodation. I have written elsewhere how moorhens will build several nests in which the chicks can roost at night. In selecting her initial nesting site, she had chosen a place where close at hand there was just sufficient material for roosting nests, and that I believe was the reason why she ordered her mate to fetch her immediate requirements from the other side of the river.

Though the heron is often rated second only to pike as an enemy of trout, I should hate to think I had seen the last one. Even though these birds destroy and damage many fish during the spawning season, there is something about a heron which compels my admiration. He is a worthy opponent, an equal match for even the most experienced stalker, and I truly believe that if one can stalk to within shot-gun range of a heron one can also stalk any other living creature. The eyes of the heron are no less keen than those of the hawk and the eagle, and big and ungainly though he may appear in flight, he nevertheless moves at a considerable speed. And when finally he alights after his graceful glide, his form takes on a symmetry which at once converts him into a creature of beauty, tensed and alive to everything around him.

He is a hunter and a stalker himself for, indeed, his living depends on his quick eyesight and his ability to make an approach to the most timid of his prey in a manner which allays all suspicion. I have seen his lazy flapping form and heard his raucous squawk at all hours of day and night, for he feeds both in daylight and in darkness. His grey-blue shape is a familiar one to me as he stands on the riverside, merged and tensed like a stalk of dead willow herb, with eyes strained to see a movement that will betray the presence of a frog, fish, vole or shrew.

Then there are his favourite stances in the valley, where he perches on a long, dead branch of a tree, perched for all to see, yet looking like part of the very branch he is on, and always at a point where he can command a view of the countryside around him. There he may sit for hours, crouched up like a moping hen, but for all his apparent lethargy and though he may look like a half dead fowl, he is very much alive. The slightest movement of a fish or a frog a hundred yards away is sufficient to attract his attention, while the clumsy movements of a stalking human are noticed almost before the stalk has begun.

I know how to kill herons, and when the need to do so arises, I destroy them, for there are times when their activities must be stopped. I find they are less cautious at early morning and late evening and, if one has the patience to wait in the line of their flight to or from a feeding place, results can be obtained. But I get little pleasure in seeing a bundle of legs and wings tumbling lifeless from the sky, or in seeing the tall grey shape slump into a heap after the impact of a charge of shot, or a rifle bullet. To me the heron is one of Nature's fishers—a fisher indeed, patient, observant, silent and sure.

Though he does certain damage in a trout fishery, he should not be condemned out of hand, for where the fishery is a natural one, he also does quite a lot of good. The occasional heron is nothing to worry about. In his stalk along the river bank, or from his lofty perch in the valley, it is not only the movement of trout that catches his eye and his needle-sharp beak is just as ready to deal a death blow to a pike or eel, to a roach or dace, bullhead or loach, all of which in one way and another are enemies of trout. He is also useful in destroying voles which, though not damaging to trout, can be very destructive to river banks and can cause endless work in repairs.

The shooting of one heron gave me much food for thought. For the second morning in succession it rose from the shallow below the bridge and, when I was about a hundred yards away, winged its way up the valley with a raucous squawk. It was a morning in early October and the sun was rising after a dark night of slight frost. I knew, as the heron knew, that many young trout had left the deep water in the dusk of the previous evening and that they had spent the hours of darkness sporting and feeding on the gravelly shallows. The older fish, too, had ranged far during the night but they, knowing full well the habits of the heron, had moved back at dawn to the deeper water and security.

"To-morrow," I thought, as I watched the bird disappear beyond the skyline, "I shall be waiting to give you something harder to digest than the young trout on that shallow."

The following morning I rose before light and went to a hide I had marked on the previous day. This was in a ditch behind a bush with a fair view of the whole of the shallow. I had my target rifle with me, a weapon of great accuracy. I made myself comfortable and waited.

Gradually the day dawned and as the first rays of the sun came over the distant hills I heard, far up the valley, the unmistakable call of the male heron. Then it was repeated, and this time much nearer. A few moments later a thin, dark line in the sky gradually took shape as the heron came straight as an arrow to the shallow. The great wings stopped beating and bent into graceful curves and, with a slow plane down, the bird alighted in the centre of the river not twenty yards from where I was hiding. I had intended to fire the moment he landed, but unfortunately part of the bush behind which I was hiding prevented me from getting a shot without moving.

Herons are possessed of extremely quick sight and acute hearing. They are particularly suspicious on alighting, and ears and eyes are alert to catch the slightest sound or movement. Not until they are satisfied as to their safety do they begin to fish. I knew that I could not move, however gently, until the bird's thoughts were on his breakfast. So I stayed still, feeling sure that in the course of his quest he would at length give me a clear shot, or at least become so absorbed in fishing that, like many anglers I know, no other thought would enter his brain.

He began, therefore, in the accustomed way, looking round and listening. I was afraid that by some uncanny instinct he knew that danger was near. He seemed to look straight into my eyes; then a movement in the water attracted his attention. The head and long neck were slowly extended and held rigid. Then, quick as a flash, down went the beak into the water which was about eight inches deep, and up it came with a small trout. This he held high in the air and swallowed without moving his position.

I was anxious to bring matters to a conclusion before more trout were lost and as the bird's attention was now on fishing, I moved a few inches, slowly and gently. As if the heron knew what was in my mind he made a stride which took him still farther behind the bush, and paused again with neck outstretched. For what seemed like several minutes he held this position and then down went the beak again into the water, to reappear with a fish of perhaps 7 in., held crosswise. The bird lifted his head to swallow the fish, but with a wriggle it slipped from his beak and back into the water. He made a quick slash at it but missed.

With head moving quickly from side to side, he searched the water for his lost prey. Suddenly he gave a quick stride forward into a clear space in front of me, but I had become so interested—in imagination I was catching fish with him— that I was too fascinated to shoot. Another lightning stroke and up came the beak with a fish of similar size to the previous one, held in a like manner.

Then followed a most interesting incident. The heron

knew for some reason that his luck was not in and he was
not going to give his latest capture any chance. Keeping a
firm hold on the trout, he started to walk towards the
opposite bank. Still interested, I held my fire. He reached
the bank, climbed up it and stopped at a point some five or
six feet from the edge of the river, having been in my full
view all the time. He then lifted his head to swallow the
fish.

He was a beautiful sight as he stood with his half-opened
beak pointing to the sky, the morning sun showing to
advantage the light-coloured feathers of his breast and
reflecting from the scales of the fish; but I had my duty to
do. He was killing trout I had spent two years in protecting,
so, hardening my heart, I squeezed the trigger. The heron's
head jerked as the bullet struck his breast, and the fish,
which would have been alive, may have been flung into the
river. It was not lying near the bird.

So far I have made little mention of the kingfisher.
Perhaps this is because the appearance of this beautiful
little bird is like a flash of thought across one's memory,
something which never stays long enough to be fully
appreciated. Perhaps, being a river keeper, I should have
no tender feelings concerning kingfishers, for though they
feed mostly on minnows, there are times, especially in late
summer and early autumn, when they can be destructive
and a menace where there are young trout. Yet even so, I
can find it in my heart to forgive them their trespasses
from the path of virtue, for throughout the years kingfishers have provided me with much enjoyment. I feel there
can be few who would not sacrifice a few trout to have
constant views of these lovely little birds as they speed over
meadow or stream.

Kingfishers are not as rare as many people imagine.
There are always some to be seen on the Upper Avon and
I often see them on other rivers about the country. But
strangely enough, though, as far as I know, no one about
here kills them, or interferes with their nests, the population
never seems to increase to large numbers. It is rare to see

more than a pair during winter and spring, in any given reach of water. I think they are like robins in that each bird or pair of birds have their own territory, their own nesting site and hunting ground, and that when the young are capable of fending for themselves, they are chased away to find a territory of their own.

Early one August I had a most enjoyable hour watching a hen kingfisher instructing her family of five how to fly and how to catch minnows. When first I saw them, all six were seated close together on a branch, with heads bobbing up and down in the way that is so characteristic of the species. The branch reached out over the water, and they were sitting about a yard away from where the nest was situated in a hollow in the tree trunk. Below them six or seven feet away, the clear, shallow water rippled over the gravel and scattered everywhere about it were minnows.

Suddenly, without any apparent warning, the mother bird flew up-stream, and then soared away up into the air for more than fifty feet. High above the tops of several trees she went, to make a circuit over the shallow, before returning to rest on the branch. To me this seemed very unusual, for the general habit is for kingfishers to fly close to the ground, or water, with a very even horizontal flight. I was therefore more than surprised when one of the young birds did exactly the same thing and then, one after the other, each of the remainder repeated the performance. Apparently they enjoyed it, for as they mounted higher and higher in the air, so they gave shrill whistles in answer to the mother, who repeatedly called them. These flights were continued time after time. As one returned to the branch, so another left it and never once were two in the air at the same time. It seemed to me, in view of what followed, that this flight was one of reconnaissance, to survey conditions on the shallows. Suddenly the mother gave an extra loud whistle and down she dived into the water to reappear at once with a minnow held crosswise in her beak. This she swallowed in the air while returning to the branch. As if her appearance beside them had been a signal, two of the young birds followed the example she had set. Down they both dived. There followed a disturbance of the water, as they struggled to rise again, but neither had a minnow. Again the mother dived, but unsuccessfully, and she reappeared, to hover like a hawk some four feet above the water. This hovering interested me exceedingly and reminded me of the humming-bird moth poised just outside a flower, with long proboscis ready to extract the nectar. The quick-beating wings caught the light from the evening sun and it seemed that all colours of the rainbow were merged into one dazzling blur. Once more she dived and this time got a minnow and came to the branch. As all the young ones clustered around her, I moved. I had been fascinated, and in my eagerness to see more, the mother bird caught sight of me. In a moment there was confusion. With shrill peeps the whole family moved as one, and all sped away down-stream.

Another incident in the life of kingfishers was almost tragic. I was about to cross a plank bridge which spanned a narrow brook just down-stream of an archway, when I heard the sound of splashing and fluttering. So, creeping to the edge, I looked down into the water five feet below. There, on the water, were two male kingfishers which at first glance appeared to be fighting. I froze, and watched, but as the struggle continued it became increasingly obvious that instead of attacking each other, the two birds were endeavouring to separate. They were locked together, for, in each case, a part of the beak was thrust down the throat of the other. One had the top part of his beak inserted, the other the lower part.

They were unaware of my presence and I watched for several minutes as they fluttered this way and that. Sometimes the struggle was violent and at others they remained almost motionless on the surface with outspread wings. I had with me a landing-net, as I was going fishing. Thinking to catch them to see exactly what was the trouble and to release them from each other, I moved forward on to the plank and bent with the net towards them. Only then did they realise I was there, and in the terrific struggle and panic of trying to escape from me, they came apart. Both were exhausted and it was some moments before they recovered sufficiently to be able to fly away.

It is difficult indeed to say what actually had happened. Perhaps they had been fighting and eventually one would have killed the other. I do not know. I have never heard that they attack or kill each other, but in view of the fact that the population seldom increases in this valley, there is the possibility that they do. The only other solution I can think of, is that they met head-on as they were speeding in the dim light beneath the arch—that, as they met, both opened their beaks in surprise and so transfixed each other with their long beaks. The beaks, being serrated, had retained their hold like fish hooks, despite the struggles of the birds to free them, and I firmly believe, had I not surprised them into making an additional effort, that both would have died.

Often I think how exasperating it must be for fish to watch the activities of swifts, swallows, martins and pied fly-catchers without being able to do a thing about it. These birds do take a great toll of the river flies. Millions of duns are killed by them each year and when you consider that approximately half of these insects are females each carrying anything from 2000 to 5000 eggs, and think that these eggs could possibly be laid to produce larvæ, nymphs and eventually a generation of flies, the loss to a trout fishery is enormous.

But there is nothing we can do to prevent this loss by birds, even if we wanted to. It is not their fault that we have meagre stocks of river flies; the fault is ours entirely. Our waters, if perfectly pure and natural, could produce sufficient insects for both fish and birds, for how often in my early days have I seen both birds and fish gorged to repletion— glutted until they have been forced to retire, and allow the river creatures to hatch and fly away unmolested. Besides, we must think of the hundreds of insects, other than river flies, that are taken by birds. Many of these, to us, are pests and, in reducing their numbers, the birds do us great service.

I think of swallows and swifts, the fly-catchers and martins, in much the same way as, in years past, the farmers considered rooks. As a boy I recollect overhearing a farmer giving orders to his carter regarding the sowing of a field of wheat.

"Much shall I put in, maister?" enquired the carter.

"Oh," came the reply. "Thee low a extry couple a zacks fer tha rooks, thay be zure ta want zum. Thay eats a main lot a pists durin' rest a tha yer, s'now Willem, en wativer we da do, thay'll ave zum carn, en I doant begrudge it to 'em."

I thought that was a very human way of looking at it. It has often occurred to me that in endeavouring to increase the fly life of the river, the swallows, swifts and martins are "my rooks" taking their inevitable toll.

It is fascinating to watch the speed and dexterity with which these fly-catching birds take their prey, and to me it never ceases to be a marvel how they avoid collision with one another. No fisherman's book can be complete without some mention of them, for indeed they form part of a fisherman's companions at the riverside. Around them many stories have been woven. There is a very old but true country saying which goes like this:—

> *When high in the sky swallows gather and fly,*
> *Rest assured for a while, the weather be dry;*
> *Low o'er the land and close to the water,*
> *Hasten poor insects, escape from the slaughter.*

However much they hasten, poor insects, once they have been sighted by those keen eyes, there is no escape. These birds know when to expect hatches of flies, and when they are speeding low over the water it is often a sign that duns are appearing. Occasionally, the birds will take flies from where they are fluttering on the surface, and several times I have seen amusing incidents when a trout and bird have collided, both after the same fly. Which of the two is more surprised it would be difficult to say, but what astonishes me is that a trout is not frightened and will continue to feed, even though a bird has just taken a fly almost from between his jaws.

At times swallows, swifts and martins will dip to the water to drink and to wash, doing both while on the wing. This skimming of the surface for drinking and washing is sometimes mistakenly thought to be the birds taking insects from the surface. A careful watch will reveal that several dips are made in the same area in succession and an absence of hatching flies makes it evident that they are not feeding.

With birds (and in this I include many species other than those I have mentioned) there is one thing which, to me, is very strange. They feed almost exclusively on hatching flies, on duns. It is uncanny how they will weave through a horde of dancing spinners, whether male or female, over land or water, and take individual duns. Fish have no such discrimination where the female insects are concerned, and will take both dun and spinner. Perhaps to birds there is little nutrient in the perfect insect, but I think it is one of the arrangements of Nature, and that fully developed ephemeral flies are a food forbidden to creatures on the wing.

When a boy, I can remember my father talking about little black river mice, but though I was forever about the riverside and adjacent streams and brooks, it was not until in my late teens that I have any clear recollection of seeing these little mammals and in finding out about them. I discovered then, that, instead of being the little black mice of which my father had spoken, they were water shrews.

The head keeper on the fishery where I was at that time knew quite a lot about them and of their habits. We had numerous trout fry in some narrow stews and it was explained to me that these little animals were taking a toll of them. Would I take a gun and watch the banks during the evenings. I was quite agreeable to doing this, and, not knowing quite what to expect, I sat down midway along one of the stews and waited. "Shoot at anything that looks like a little shadow moving near the water's edge," I had been told, and in less than half an hour the order was obeyed.

For a while I thought I had killed a little mole. The resemblance was very striking. There was the plump little body and blackish-blue silky fur, and this, together with the long-pointed nose and tiny eyes, gave me cause to wonder if I had killed the right animal. Quickly I found the head keeper and showed him the victim.

"Oh, you managed to get one, then," he greeted me. "I heard you shoot."

"It is a shrew, then, after all," I replied. "I thought I had made a mistake and got a mole."

He laughed. "You won't find a little mole with a white belly and a tail as long as that, Frank. That's a water shrew all right, the little beggars are fond of trout fry. Try again, you might get another. They are doing a lot of damage out there."

But I did not get another and as I write now I am rather pleased about it. I have seen many other water shrews since then and have had chances to study individuals at close quarters. I have found out, too, that the head keeper was right in saying these little mammals are fond of trout fry— they are—and of alevins and eggs, too. Some of my best studies have been in my trout hatchery, for each year before the war, at much about the same time in January, a shrew would put in an appearance. The interior of the hatchery is darkened. This probably accounts for the appearance of shrews in the daytime as I believe they are naturally nocturnal animals.

The first one or two shrews did not attack any of the trout eggs laid down in trays for incubation, but were quite content to take dead eggs which, carelessly, I had let fall to a beam near the level of the water over which the hatchery was suspended. Often I sat and watched them as they gathered up the eggs and ate them. Then one year, my assistant complained of losses of alevins in one tray, for which he could not account. I suspected shrews and told him to cover all the trays securely. Losses then ceased. Then one morning a month later, when the little fish had almost reached the fry stage, I made a visit to the hatchery.

As I opened the door a small shadow appeared from beneath a cover of one of the trays, but before I could be sure what it was, it was gone. Closing the door, I waited. It was but a moment or two when the shadow appeared again and I could see it was a water shrew. Without a pause it crept beneath the loose cover of one of the trays and then I heard a disturbance of the water as it dived after the trout fry. It dived and reappeared a dozen times or more, and each time with a mouthful of trout fry. These were eaten at lightning speed as it sat on the edge of the tray. How many it would have taken if left undisturbed I cannot say, but even for the sake of nature study I could not let it continue to feed on such a delicacy. I made a move to catch it but did not succeed. Several hundreds of fry were missing from that one tray.

The water in the tray containing the trout fry was three inches deep and each time the shrew had to submerge to get its food. Yet, on each occasion I saw it on the edge of the tray, the little animal looked to be perfectly dry, with no fur displaced or bedraggled.

But it would appear that water shrews will take fish much larger than young trout fry. One morning in early January I was walking gently along a spring-fed stream when a water shrew appeared from a hole in the bank. Running down a bit of a slope it entered the water and immediately submerged. I had a clear view as it moved, like a giant beetle, along the bed of the stream beneath about four inches of water, until it came to a pool containing a few big stones. Here the water was deeper—perhaps eight inches deep. For a moment or two the little mammal appeared to be struggling hard to get beneath one of the big stones, and then, moving to one side, it disappeared under it. A pause, then out it came to move frantically, still under water, towards the place on the bank where first I had seen it. I could see it had a fish of some sort in its mouth, but apparently the bank was too distant to be reached without surfacing. About a foot from the outside it came up, and then swam quickly to the land.

Here the shrew paused for a moment and I could see that the fish was a stone loach of about 3 in. long, and was held by the tail. The loach threshed madly about and it appeared as though the shrew had some difficulty in holding it. But, without moving its grip the animal went backwards up the bank dragging the fish along and then entered the hole of a water vole. Though I waited for a long time I did not see it again.

Another study of a shrew feeding was made in June. Probably for some natural reason of which I had no knowledge, toads had spawned in exceptional numbers near and in the river. As a result, by early June, many thousands of tadpoles were in every ditch, brook and stream. I was interestedly watching some in one of the streams as they tried to cluster beneath some leaves and other debris. So intent was I on this subject that I did not see the shrew enter the water. I was more than surprised to see a flurry in the leaves and to see it busy amongst the tadpoles. Only for a moment did it remain, then, returning under water to the bank, it quickly climbed out and rushed into a hole. About five seconds later it reappeared, and this time I watched closely.

With scarcely a ripple it entered the water and was quickly searching in the leaves at the bottom. As it returned and surfaced it paused long enough for me to see the tails of several tadpoles hanging from its mouth. Then into the hole it went again. This was repeated time after time and an area of perhaps ten square yards was searched. Each time it caught tadpoles. I had quite made up my mind that it had a nest of young ones in the holes and was feeding them, and then out it came, swam across the surface of the stream to run at least twenty yards along the edge of the bank, to disappear for good into another hole. Though I dug into the bank and followed the first hole to its end, I could find no trace of a nest or of young, and of the numerous tadpoles I had seen taken in there was not a trace. The shrew must have eaten them all.

Under water the shrews appear to be whitish because of

the air trapped in the velvety fur. Their underwater movements are always very rapid and remind me of waterboatmen, or some of the large aquatic beetles. They make a nest in a hole in the bank above water level in which to have their young in spring. The nest is similar to that of the water vole, but smaller, and usually made of dried grass, rushes and occasional dry leaves. A fully grown male may have a body 3 in. long with what I call a three-quarter-length tail. Some water shrews have white ears, but whether this is just a feature of the male I cannot say. I have caught females without white ears, but not enough of either sex to be certain.

Their favourite places are springheads, watercress beds and spring-fed streams, for here it is that a plentiful food supply is to be found. Though mostly they choose to move under water by running along the bed of stream or river, as surface swimmers they are perfect. Many times I have seen one rapidly cross a stream with scarcely a ripple to mark its progress—just a tiny shadow speeding over the surface—to make me wonder if it is swimming or flying, or if indeed I have seen anything at all.

It takes a fall of snow to make one fully realise that it is winter, and nowhere can the transformation be more pronounced than in a river valley. The great mantle of whiteness shows only too well the time of the year and exposes to view many things we would like to remain hidden. To me the sight is depressing. There is the river, cold and forbidding, moving slowly down the lower parts of the valley towards the sea, like a ribbon of black polished steel. Here and there a pond or a pool shows vividly amongst the whiteness, while the brooks and streams stand out starkly as though one has spilled a bottle of ink over a sheet of paper. About the meadows dried willow herb and meadowsweet stand side by side with docks and thistles, comfrey and figwort, rearing like dead sticks above where the rushes and grass are muffled in the powdery flakes. Here and there the trees of the valley, with gaunt spreading limbs, reach their bare branches towards the leaden sky, as though in supplication.

215

Over everything is silence—a silence so great that one feels alone amongst a Nature that is dead. It is the silence that I hate most. Everything is muffled and held tightly by the grip of snow. No longer can I hear the little rustlings of the wind in the dried grass and rushes of the riverside, or as it whispers its way through the withy beds. The birds are all hushed as though overwhelmed by the calamity, while the river and stream creep silently on their way, as if conserving their gurgling and chuckling for the thaw.

But though the valley is silent and unfriendly the riverside can be interesting after a night of snow. On the banks of river and stream many stories can be read of the hunters and their prey. In imagination I can follow them on their journeys. Otters seem to glory in a snowfall. I see the drag marks of their tails where they have crossed from river to pond or to brook and where they have rolled and played in the snow and ploughed furrows through the drifts.

Sometimes I see where all four footprints together show they have loped along like great hares to cover four or five feet at a bound. Then a straight line of pad marks show where a dog-fox or vixen has been on the prowl, and then smaller prints, something like those of a tiny otter, tell me stoats or weasels have been hunting. Here and there go the marks of a rat or vole, or perhaps the little prints of a fieldmouse or shrew lead into a tuffet beneath the snow.

Near the river's edge are criss-crossed the prints of moorhens, and a wet, dabbled patch shows where mallard have rested. And I see where the heron has stalked along, his big three-toed imprints leading into the snow-covered rushes at the riverside. I imagine him as he pauses with head outstretched and beak ready to drive downwards at the slightest movement of vole, mouse or shrew. And as I plough my way along so I see some of the birds of river and riverside. But no longer are they friendly. The moorhen scuttles away and plunges headlong into the nearest cover with a flurry of snow, while the dabchick ducks quickly beneath the water. Mallard rise into the air and go quacking away long before I have a chance to see them on the river, while every now and then a snipe rises from my feet to go zig-zagging up into the sky, as though I was a giant coming to destroy them all.

How different it all is. I pause for a moment and let my thoughts wander away to the time when spring will be here again. What a contrast there is in December and May—the sleep and the awakening of Nature. As I muse, my eyes catch a movement, and there, as if reading my thoughts, a robin sits on a branch. He perks his head to one side, his little red breast a vivid patch of colour amongst the blanket of whiteness, and then he whistles loud and long.

Amongst the great solitude the little tune sounds clear and strong. In it is a whisper of friendliness, and of spring. It cannot last, he seems to say. It is but the season of the year. Let us look forward to to-morrow.

217

INDEX

FRANK SAWYER: MAN OF THE RIVERSIDE
by Sidney Vines £9.95

'This is an intriguing and highly readable work – a most fitting tribute to a most remarkable man.'

Peter Lapsley – Shooting Times

'Frank Sawyer . . . lived and breathed the river. Sidney Vines' tribute to the celebrated keeper of the Avon.'

Ian Niall – Country Life

'What comes over clearly is Sawyer's passionate devotion to the river which eventually became his charge – he was one of those rare creatures: a born genius.'

Conrad Voss Bark – The Times

DAYS & NIGHTS OF GAME FISHING by W. B. Currie £9.95

'I would place this book on an accessible shelf with Negley Farson and Arthur Ransome . . . A potential classic, this elegantly written book with its apposite illustrations by Charles Jardine is ideal for any fisherman.' George Melly – Times Literary Supplement

THE EVER ROLLING STREAM by Bernard Aldrich £8.95

'Keeper of the Broadlands beat, Bernard Aldrich writes: "in fact, we mostly look after our waters as if they belong to us – although we don't object to our employers actually owning them, they are really ours". The book is his own, is relaxing and charming – just like the man himself.'

Sidney Vines – The Field

'Mr Aldrich taught Prince Charles to fish and knows every twist of the Hampshire Test – delightfully illustrated by Manuela Smith, here are more than fisherman's tales.'

Shaun Usher – Daily Mail

THE TROUT LOCHS OF SCOTLAND – A Fisherman's Guide
by Bruce Sandison

'For lovers of wild places, this book will be a key to a world they have only dreamed about.' £4.95 paperback

Trout and Salmon

BOAT FISHING FOR TROUT by Steve Parton £4.95 paperback

'Destined to become the standard reference work for the modern fly fisher'. Trout Fisherman

NEW ANGLES ON SALMON FISHING
by Capt. Philip Green R.N.

A perceptive guide that will help every fishing enthusiast toward greater success. £9.95